IMAGES
of America

FARMING IN ANNE ARUNDEL COUNTY

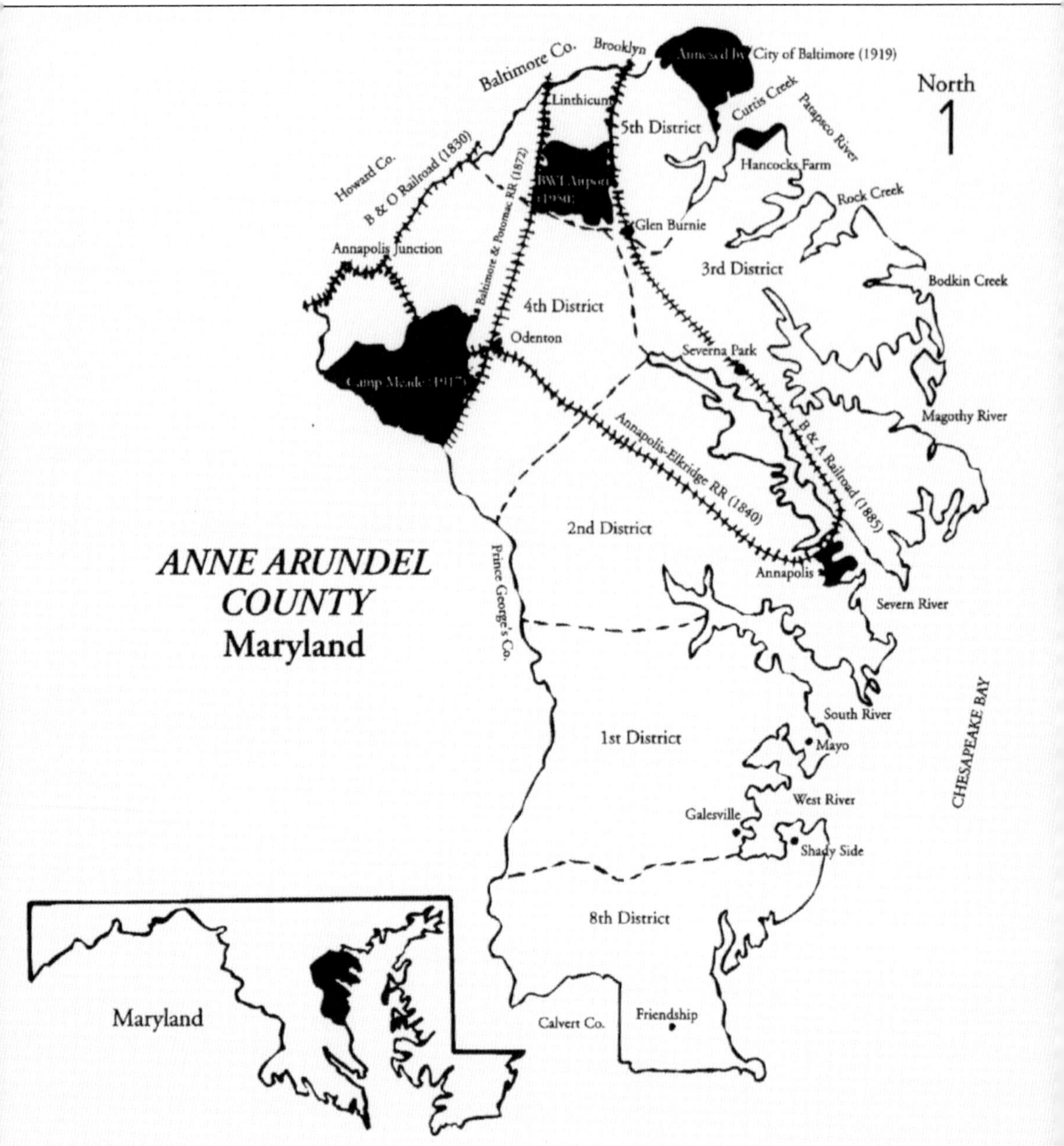

The small map shows the location of the county in Maryland, and the larger map depicts some of the towns located in the county. Baltimore City borders to the north, and Washington, DC is located to the west. (Mark Schatz and the Ann Arrundell County Historical Society.)

On the Cover: Farmers hang tobacco in a tobacco barn on a farm in southern Maryland in 1937. A hogshead used for packing tobacco is in the foreground. For centuries, this harvest scene was typical on farms throughout Anne Arundel County. (Enoch Pratt Free Library.)

IMAGES
of America

FARMING IN ANNE ARUNDEL COUNTY

Frederick H. Doepkens

ARCADIA
PUBLISHING

ISBN 978-0-7385-8756-1

Published by Arcadia Publishing
Charleston, South Carolina

Printed in the United States of America

Library of Congress Control Number: 2010943161

For all general information, please contact Arcadia Publishing:
Telephone 843-853-2070
Fax 843-853-0044
E-mail sales@arcadiapublishing.com
For customer service and orders:
Toll-Free 1-888-313-2665

Visit us on the Internet at www.arcadiapublishing.com

This book is dedicated to my parents, William P. Doepkens and Marjorie Desmarais Doepkens, in loving memory.

Contents

Acknowledgments		6
Introduction		7
1.	Tobacco Farming	9
2.	Truck Farming	51
3.	Livestock and Dairy Farming	67
4.	Grain and Forage Crops	91
5.	Horticulture and Nursery Crops	107
6.	Managing Our Natural Resources	117
7.	Educating Our Future	121
Bibliography		127

ACKNOWLEDGMENTS

Farming in Anne Arundel County is a labor of love that enabled me to reconnect with my agricultural roots and give back to the community in which I spent my childhood. With the support of Chris and Jeff Griffith of the Anne Arundel County Farm Bureau, over 2,000 photographs and histories were collected from numerous farm families and organizations.

Thanks to the following organizations, historical archives, and individuals:

"Reprinted with permission of the Baltimore Sun Media Group, All Rights Reserved" (Sun); collection of the Maryland State Archives (MSA); Anne Arundel Soil Conservation District (AASCD); Special Collections, University of Maryland Libraries–Agricultural Experiment Station Collection (UMCP); Special Collections, University of Maryland Libraries–Baltimore News American Collection "by permission of Hearst Communications, Inc., Hearst Newspaper Division" (Hearst); published with permission of the capital, Annapolis, Maryland (Capital); Mark Schatz and the Ann Arrundell County Historical Society (AACHS); the Friends of Kinder Farm Park Historical Society (Kinder); Smithsonian Environmental Research Center (SERC); Anne Arundel County Public Schools (AACPS); Enoch Pratt Free Library, Maryland's state library resource center, Baltimore, Maryland; Don Riddle Jr., Tim Hamilton, and Ann Marie Sedor of Homestead Gardens; Allen Colhoun and the Colhoun family of Ivy Neck Farm (Colhoun); William H. Gertz (Gertz); the Griffith family in Lothian (Griffith); John M. "Jack" Scible (Scible); William Mueller (Mueller); Edwin Calvert (Calvert); Emma M. Schramm (Schramm); Mildred Bottner Anderson (Anderson); Nancy and Max Covington Jr. (Covington); Oscar F. Grimes Jr. (Grimes); the Tucker family album from Margaret Chase (Chase); photographer Margie Salyers (Salyers); photographer Frederick H. Doepkens (FHD); Heather Wheatley; Joanie Surette; and other contributors who are listed throughout the book.

Unless otherwise noted, production statistics were from the *Census of Agriculture* from the National Agriculture Statistics Service, US Department of Agriculture.

I would like to thank my friend and colleague Michael Thorne for editing all of the text and Elizabeth Bray at Arcadia Publishing for guiding me through the process. A very special thanks to my family and friends for their contributions, support, encouragement, and patience throughout the entire endeavor—especially Margie Salyers, Liz Yuan, Jean Wright, Christina and Katherine Salyers, Alyssa Yuan, and Paul Ammirata.

Introduction

Agriculture is the basis of civilization. It provides all our basic needs: food, clothing, and shelter. Over the past 350 years, both American agriculture and the ingenuity of American citizens have transformed this country from a fledgling nation into a superpower. The citizens of Maryland and Anne Arundel County have played a major role in these advancements. Anne Arundel County was formed in 1650, and by 1900, it was dotted with small farming villages and summer resorts along the waterways. Located in the center of the state of Maryland, it has now become part of the quickly growing Baltimore, Washington, and Annapolis metropolitan area, where quiet pastoral landscapes are making way for the hustle of suburban life.

As we get further and further removed from our agricultural roots, it is important for us as a society to step back and reflect on the farm families that helped to shape our state and nation. This book helps to preserve for future generations the agricultural heritage of Anne Arundel County by exploring life on the family farm and the methods used to raise the livestock and grow the crops that many of us take for granted today. This photographic journey will help to put a face on American agriculture and the family farm in our local community. The chapters are divided into the types of agriculture found in the county: tobacco, truck crops, livestock, grain crops, horticulture and nursery crops, and natural resources. The final chapter examines the role of education in helping to ensure a future for the agricultural industry. In the chapter on tobacco, unless otherwise noted, the supporting facts came from *Maryland & Tobacco*, published by the Tobacco Institute.

From the founding of the colony of Maryland in the new world in 1634 to the birth of our nation, agriculture has changed and adapted to meet the needs of society. Tobacco was a dominant crop grown in Maryland since colonial days. The type of tobacco grown is oronoko, commonly referred to today as Maryland Type 32 tobacco. During colonial days, it was used as currency, and up until the year 2000, it was a major cash crop for southern Maryland farmers. Except for the introduction of mechanization and chemical fertilizers, the farming methods used to grow the crop changed very little in the past 300 years. According to the Tobacco Institute, it takes about 230 man-hours per acre to produce Maryland tobacco, and from seedbed preparation to sale of the crop takes 18 to 20 months. Due to Maryland's climate and the unique sandy soils in the southern part of the state, the tobacco grown there exhibits unique, even-burning qualities demanded by cigarette manufacturers.

Most of the crop was initially exported to England, but later, Baltimore became a center for tobacco processing. As the importance of tobacco grew in trade, so too did the demand for slaves to help raise the tobacco on the southern Maryland plantations and manors. By 1860, Maryland was producing 51 million pounds of tobacco per year. After the Civil War and the abolition of slavery, tobacco production dropped to only about 21 million pounds per year. By the end of the 20th century, Maryland was producing around 12 million pounds per year. In 2001, a state buyout program began paying farmers not to raise tobacco and to convert to raising other types of crops. This program effectively ended the 350-year tradition of Maryland tobacco production. In 2004, Anne Arundel County only produced 155,500 pounds of tobacco. The process of tobacco production is explored through this photographic journey.

With the rise and fall of tobacco production in Maryland, other types of farming began to emerge. Diversification was the key to survival for the family farm. During the last part of the 1800s and the early 1900s, many new immigrants made their way to Anne Arundel County from around the world in search of a better life. Farmers diversified into other types of crops in addition to tobacco. Truck crops, corn, wheat, soybeans, hay, and the raising of livestock became more important in our local rural economy and helped to strengthen the family farm.

In the northern part of the county, truck farming replaced tobacco as the major crop. According to Willard Mumford's book *Strawberries, Peas & Beans*, truck farming is the growing of vegetables for a commercial market. Truck crops dominated farm production in the northern part of the county from the 1870s and remained strong until the late 1930s. Many of these crops made their way to Baltimore markets. After World War II and subsequent improvements in transportation routes and refrigeration, there was a decline in production. The truck farming chapter will highlight some of the world events that also had a role to play in shaping our agricultural heritage. During World War I, there was a typhoid outbreak in the local milk supply. In response, the US Naval Academy purchased land in 1913 for a dairy farm that would provide a healthy milk supply for our military men in training at the academy. In 1997, the dairy farm closed its doors due to the increased cost of production and the presence of cheaper sources of high-quality milk in the marketplace. During its tenure, the farm provided a model for farming practices that became commonplace throughout the agricultural industry.

As we move into the 21st century, agriculture in Anne Arundel County is still adapting to changes in the local economy and pressures from society. Farmers are searching for new ways to continue family farm traditions in this ever-changing marketplace. Technology is revolutionizing the methods of farming, and new markets are providing hope for specialty crops and a strong future for agriculture. Production of horticultural and nursery crops has increased.

As the son of a diversified crops and livestock farmer, I worked on the family farm during my youth and experienced first-hand the raising of tobacco. Eventually, like so many others, I left the family farm in search of a better life. I hold close to my heart my agricultural roots and heritage, and as an agricultural educator for over 30 years, I have strived to educate the public about the role of agriculture in our lives and its importance to society. Farmers are stewards of the land and over the years have developed improved farming practices that helped increase production while conserving our natural resources and sustaining our agricultural lands.

Farming in Anne Arundel County explores my experiences growing up on a family farm as well as those of countless farm families who tilled the soil in search of a better life. These experiences are presented in historical context, creating a snapshot of agriculture in Anne Arundel County that will educate the public about agriculture—the largest industry in the state of Maryland.

One

Tobacco Farming

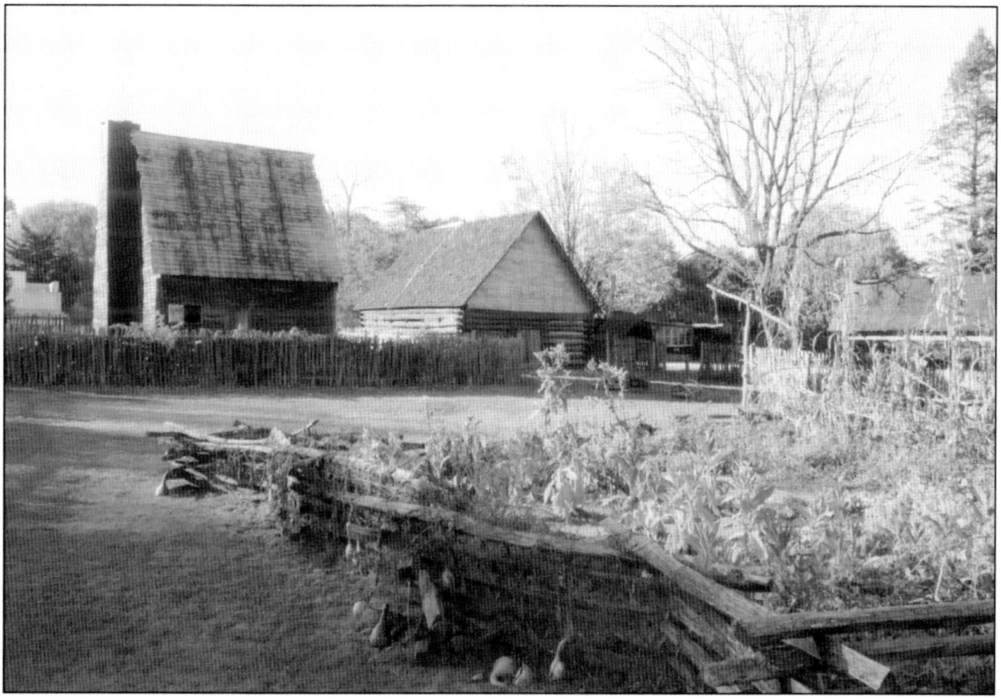

Tobacco has been a dominant crop in Maryland since colonial days. In this recreated colonial homestead on the grounds of the Historic London Town and Gardens in Edgewater, pictured in 2010, tobacco plants grow in the garden, and the house and tobacco barn are in the background. Maryland tobacco is known for its unique burning qualities due to the variety, climate, and soils of southern Maryland. (FHD.)

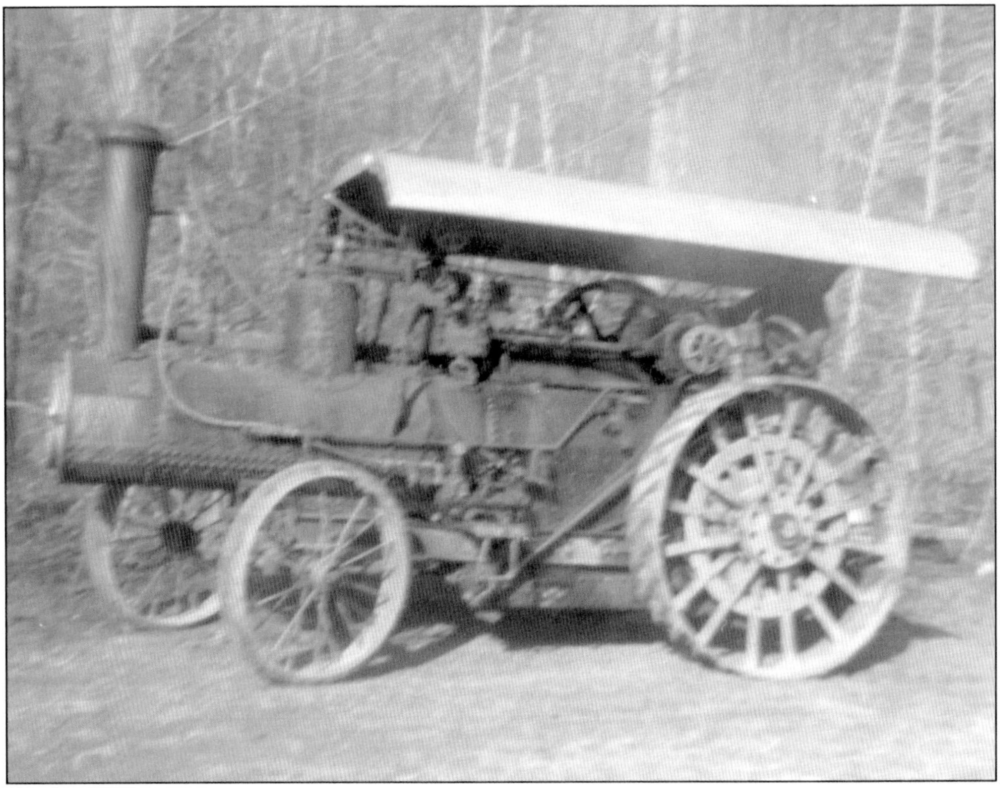

An area used for planting the tobacco beds is prepared in late September and October. It is usually an area that is near a water source, with a southerly exposure and protection from wind. This steam tractor owned by Francis M. Gasperich in Crownsville is being used to help sterilize a tobacco bed in 1966. (Gertz.)

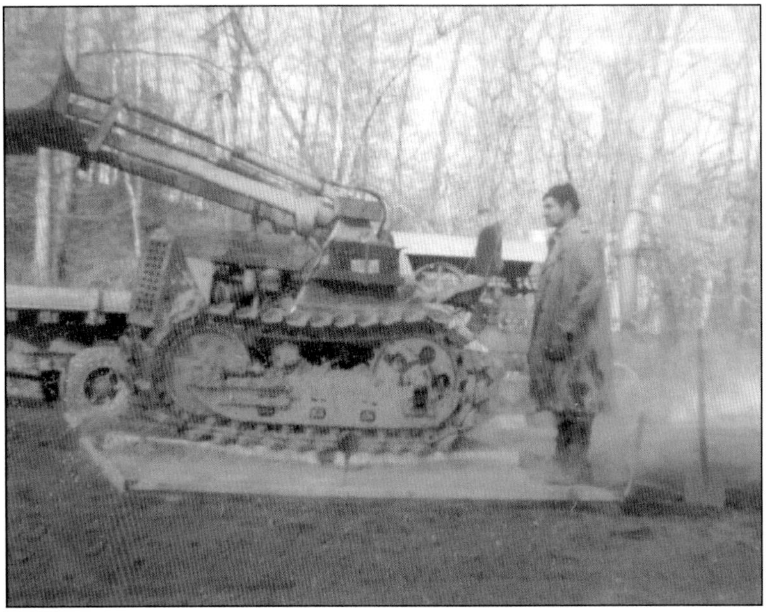

In order to steam the beds on the Gasperich farm in 1966, farmers would connect a steam hose from the steam tractor to a metal plate resting on the ground. The bulldozer would be driven on top of the plate to help hold the plate down onto the ground during the steaming process. (Gertz.)

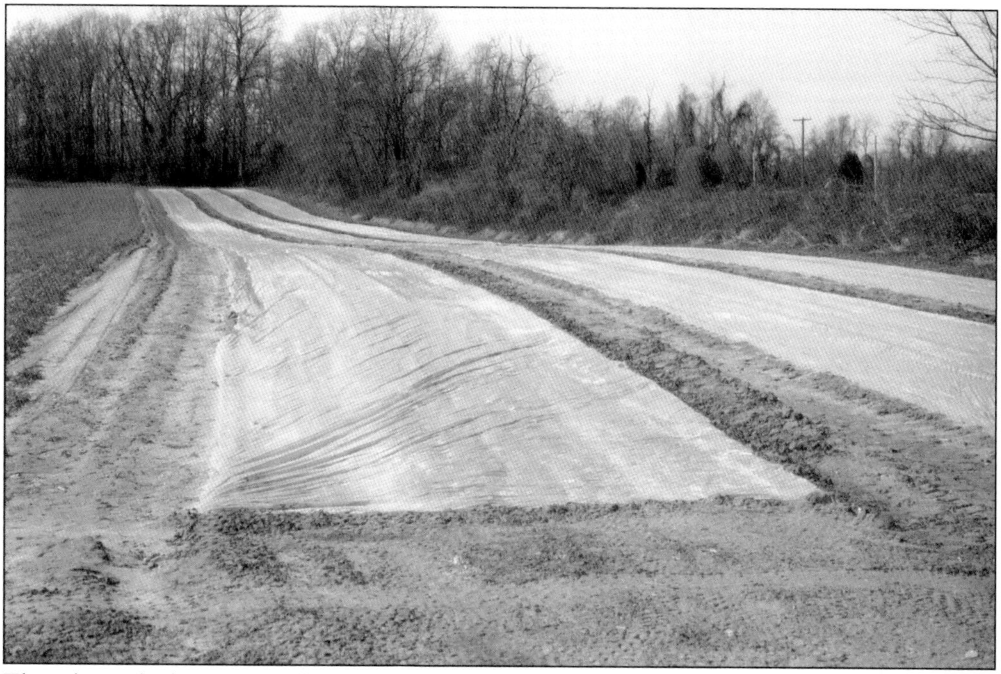
The tobacco beds are covered with plastic on the Earl Griffith farm in March 1995, and the edges are sealed in with soil to keep the ground temperature around 50 degrees. The beds are treated to prevent the growth of weeds, insects, and diseases. (Griffith.)

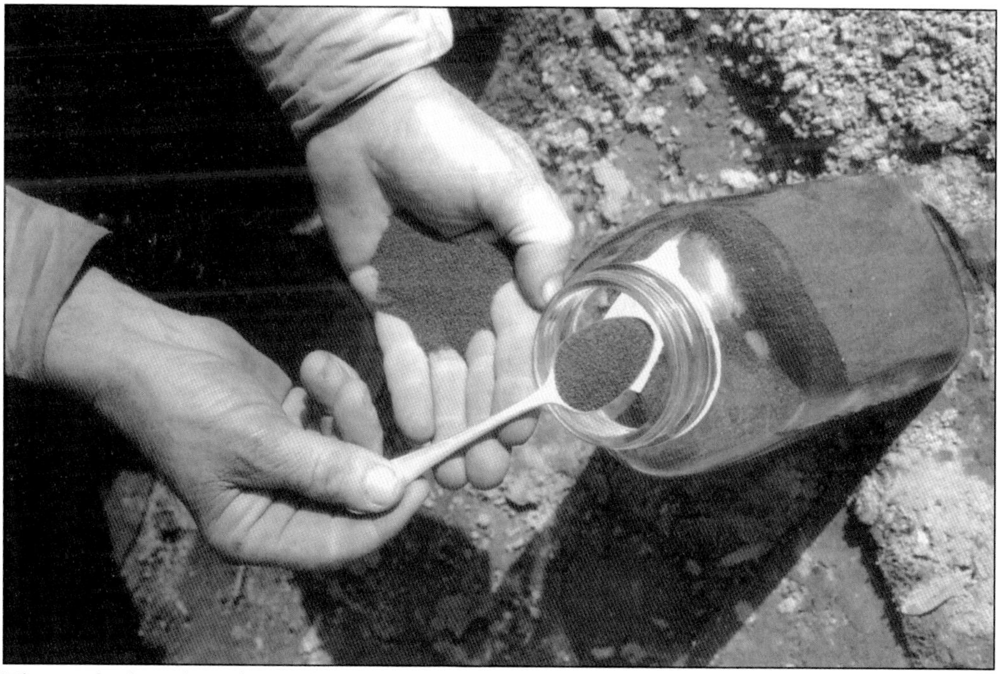
The seeds that the tobacco farmer used were collected from tobacco plants on the farm, or different varieties of Maryland Type 32 tobacco seeds could be obtained from the Maryland Tobacco Growers' Association. Seven million tobacco plants can be raised from one pound of seed. This photograph is from 1947. (Sun.)

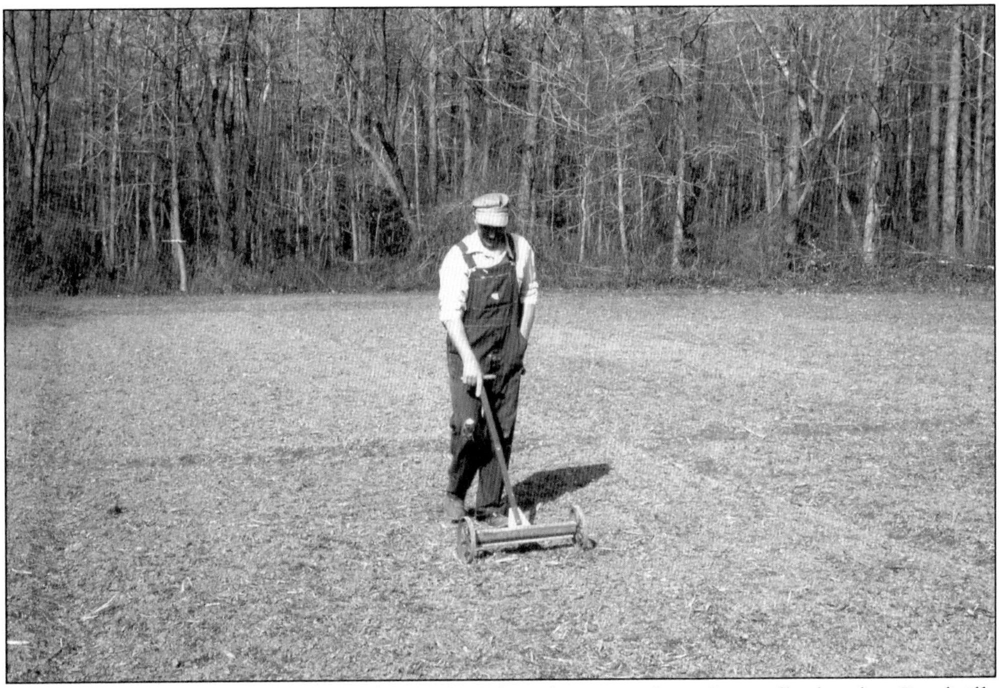

William P. Doepkens uses a drop seeder to plant the tobacco seeds in the seedbed on his Gambrills farm in March 1976. Other ways of planting the seeds include using a rotary seeder or spreading seeds on the surface of the soil by hand. After the seeds are planted, they are tamped down by foot or by the use of a roller. (FHD.)

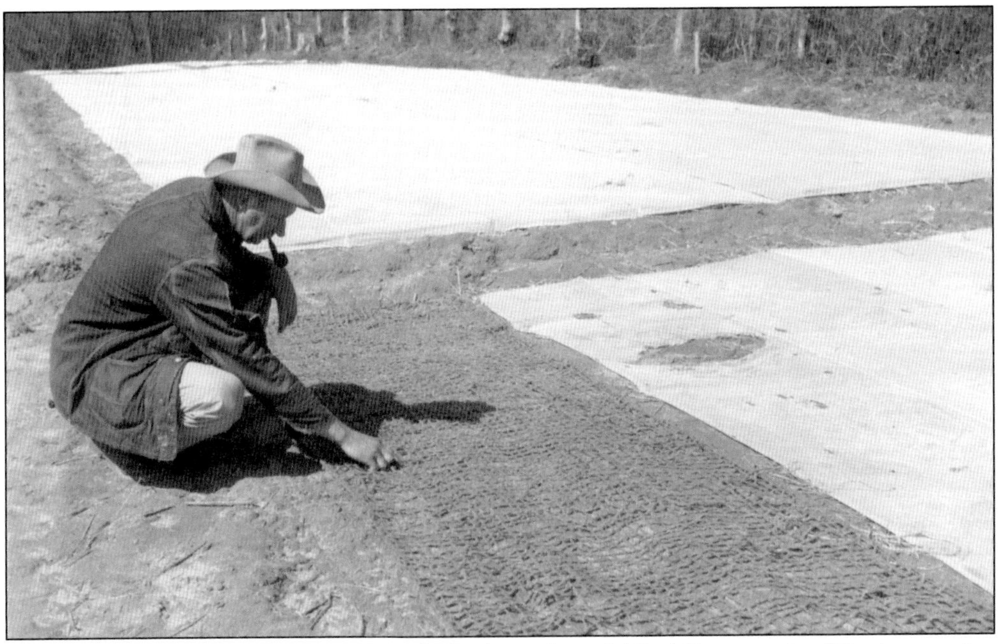

While participating in an experiment with the Anne Arundel Soil Conservation District on using jute as a cover for the tobacco beds in 1952, Charles Bassford examines his own tobacco beds. The beds are covered by cotton, which protects the plants as they begin to grow. (AASCD.)

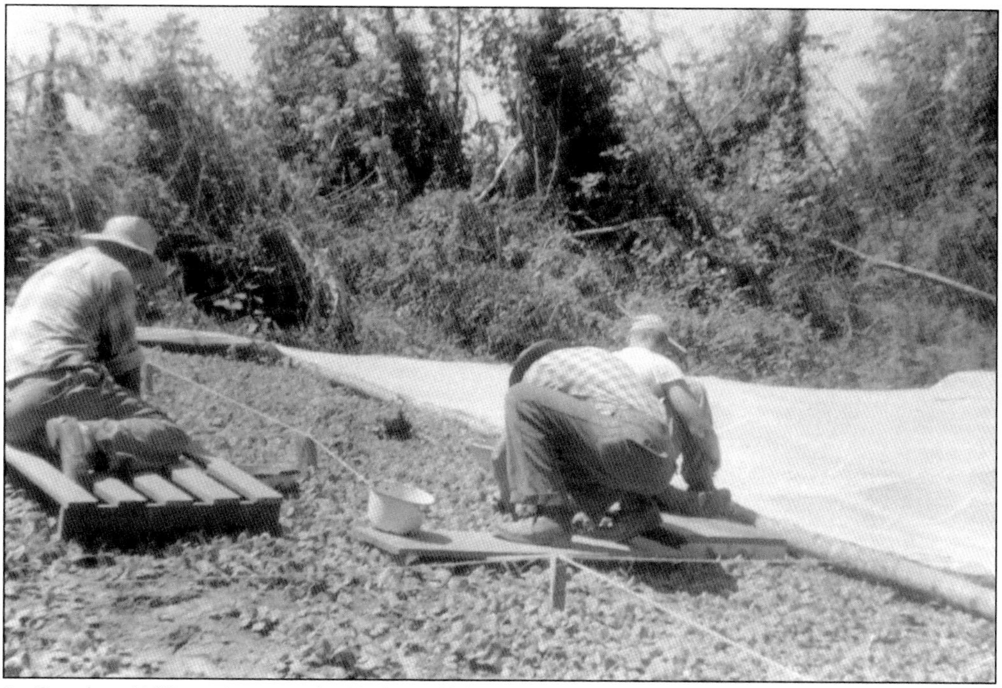

In October 1957, workers on the Earl Griffith farm in Greenock check out the newly emerging tobacco seedlings. The seeds begin to germinate around mid-April. Farmers irrigate their tobacco beds as needed to insure maximum growth of the young seedlings. (Griffith.)

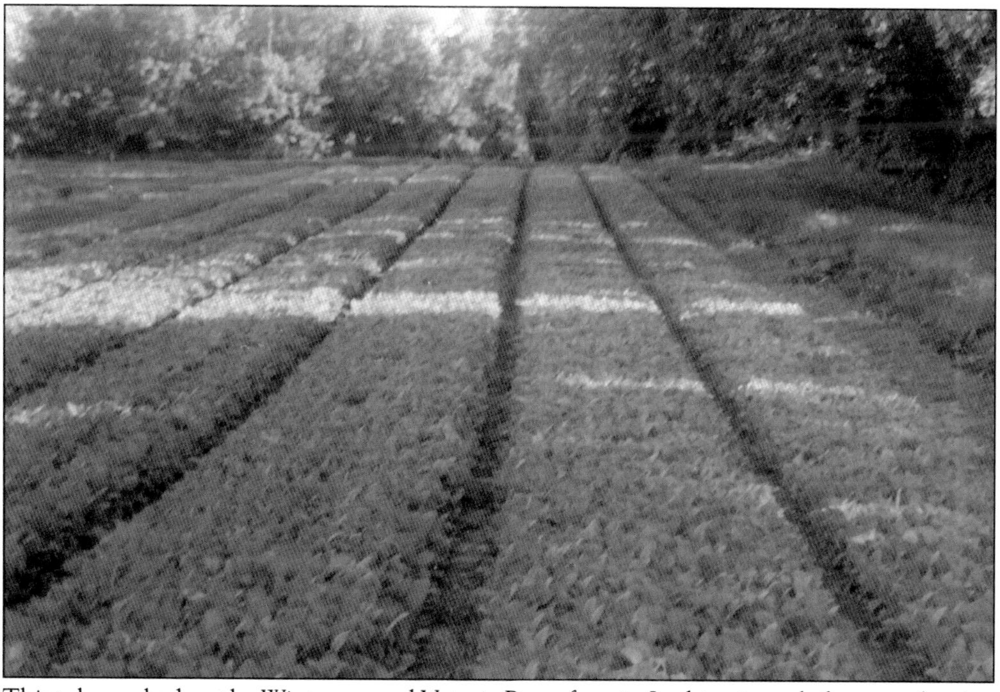

This tobacco bed on the Winterson and Veturia Prout farm in Lothian is ready for transplanting to the field in 1983. By late May and June, the tobacco plants have reached the desirable size of about 10 inches tall and are ready to be transplanted into the field. (Griffith.)

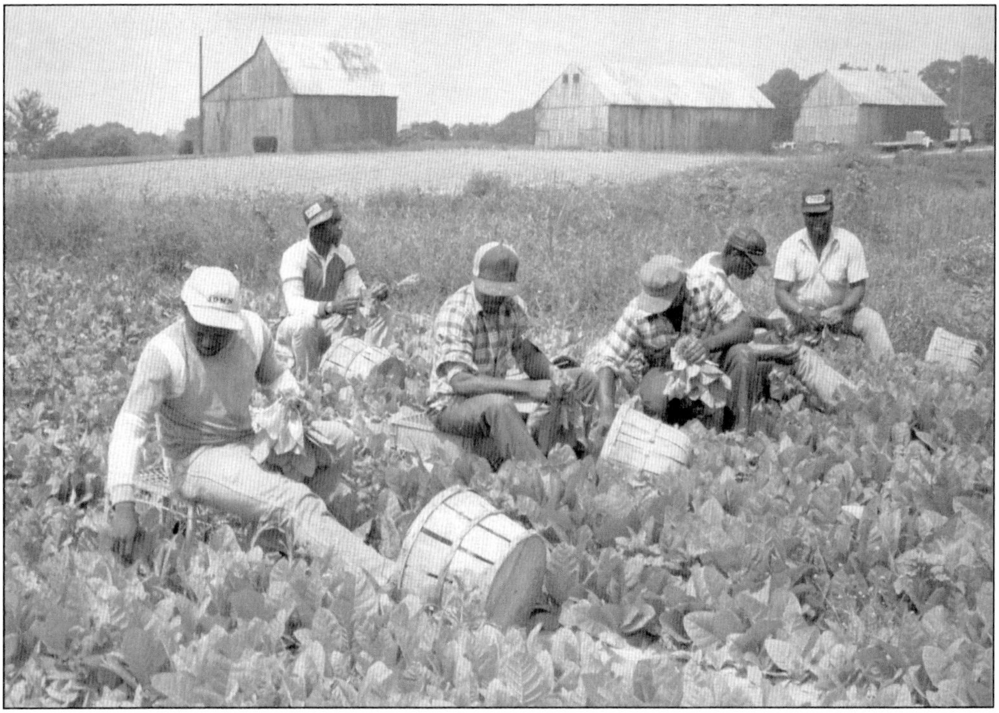

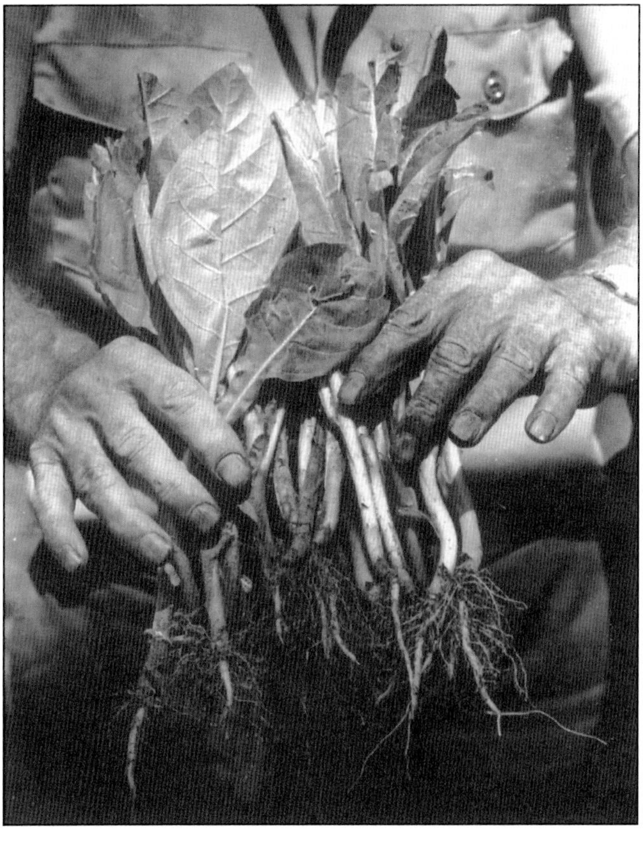

Workers on the Earl Griffith farm around 1990 pull selected plants one at a time and neatly prepare and pack them into bushel baskets. Some farmers also use boxes for packing plants to get them ready for transplanting. Plants are usually pulled in the morning, when the soil is moist, and then are transplanted into the field during the afternoon. (Griffith.)

Oscar F. Grimes Jr. holds a handful of freshly pulled tobacco plants in June 1984. The 10-inch plant should be well rooted and should have a clean stem with several well-developed leaves on the top of the plant. The plants are neatly packed into bushel baskets or boxes. (Hearst.)

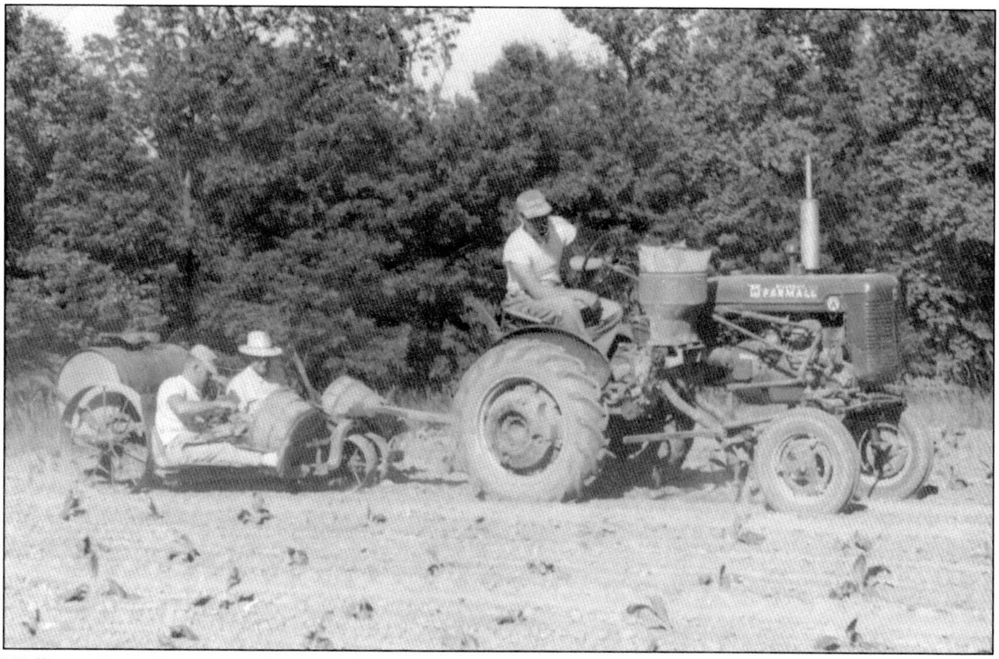

William Gertz drives a 1953 Farmall Super A tractor, which is pulling a New Idea one-row tobacco planter on his farm in Crownsville as his dad, Emil Gertz, and Herman Stevens sit on the planter and plant tobacco in 1957. This planter used to be pulled by horses but then was modified for tractor use. (Gertz.)

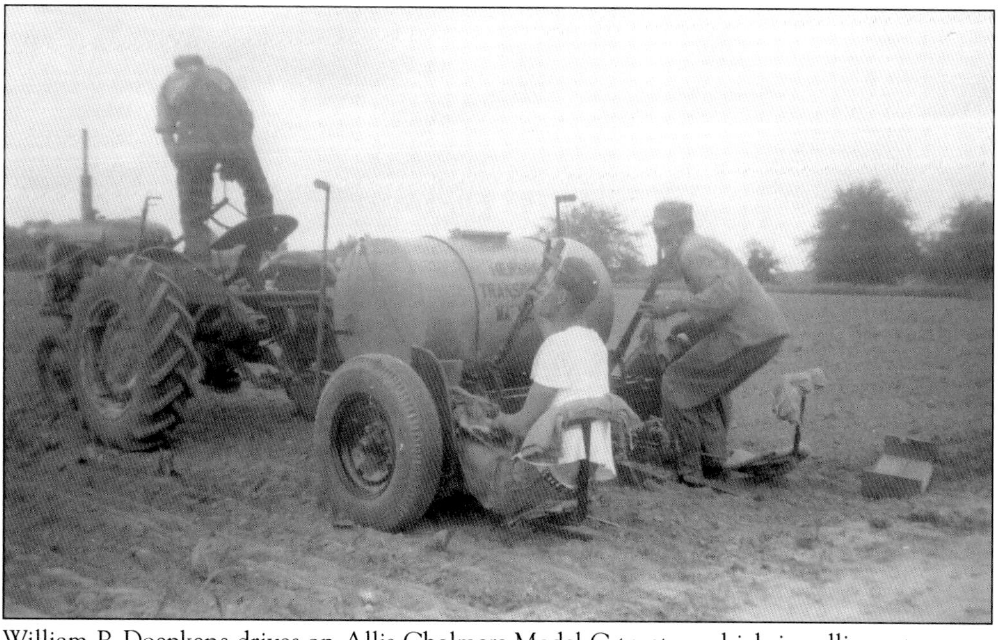

William P. Doepkens drives an Allis Chalmers Model C tractor, which is pulling a two-row, two-person tobacco planter on the Henry Doepkens farm around 1950. The planters, John M. "Jack" Scible (left) and John Johnson (right), would load tobacco plants onto a long wooden tray that would sit in their lap. Plants would be pulled from this tray and then inserted into the open furrow for planting. (Scible.)

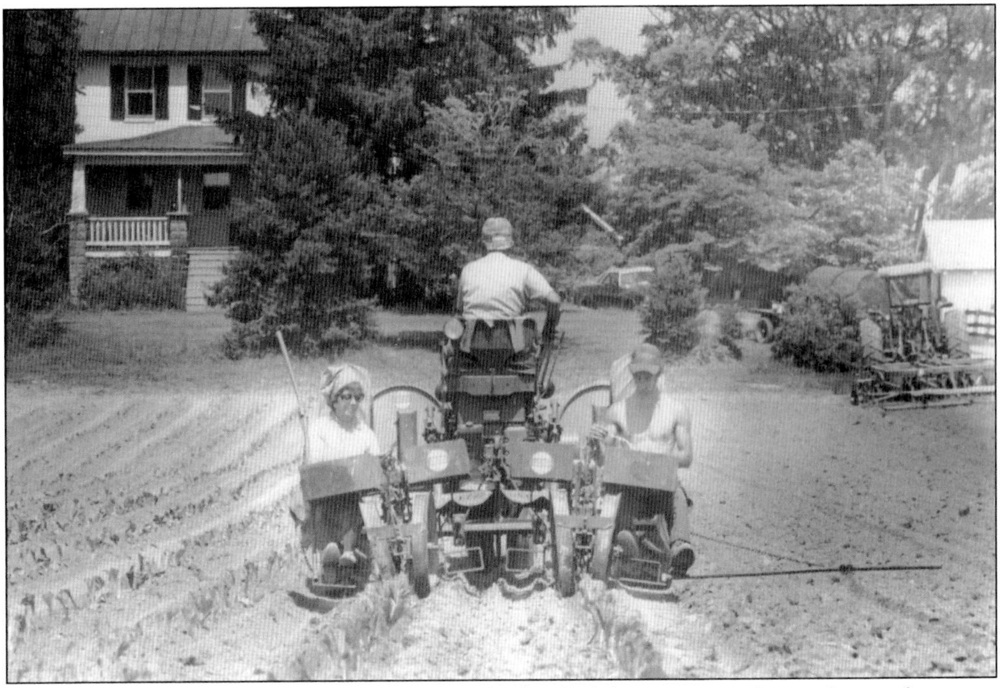

Oscar F. Grimes Jr. drives a tractor attached to a hydraulic-lift two-row mechanical transplanter on his Davidsonville farm on Birdsville Road in June 1984. Water barrels are attached to the tractor. The planters, Jean Grimes (wife) and William Grimes (son), plant their own rows. The water to fill the tanks is obtained from ponds, streams, or wells located on the farm. (Hearst.)

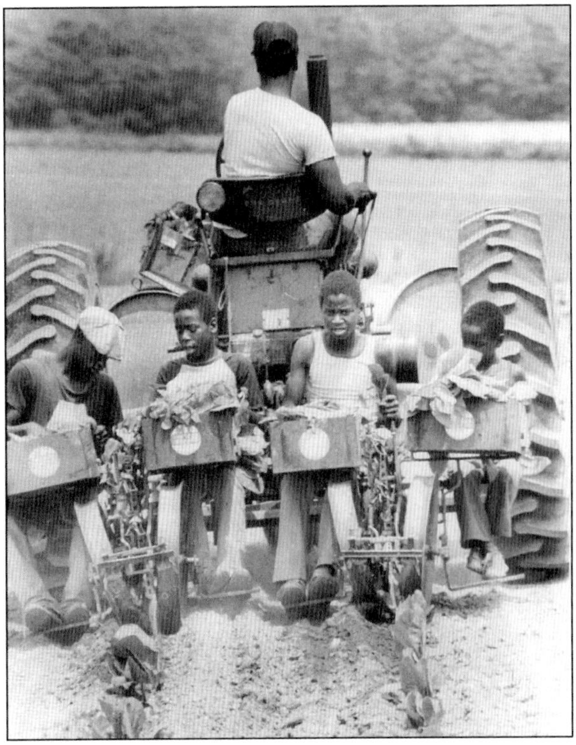

On the William E. Zang farm on Muddy Creek Road in Harwood in 1983, Daniel Randall drives a tractor. The planters (from left to right), Milton Tongue and Danny, Tracy, and Corey Randall, place the young plants onto small trays attached to a revolving wheel that inserts the plants into the ground. The 100-acre Zang farm now raises peaches and vegetable crops, which are sold at its farm stand. (Capital.)

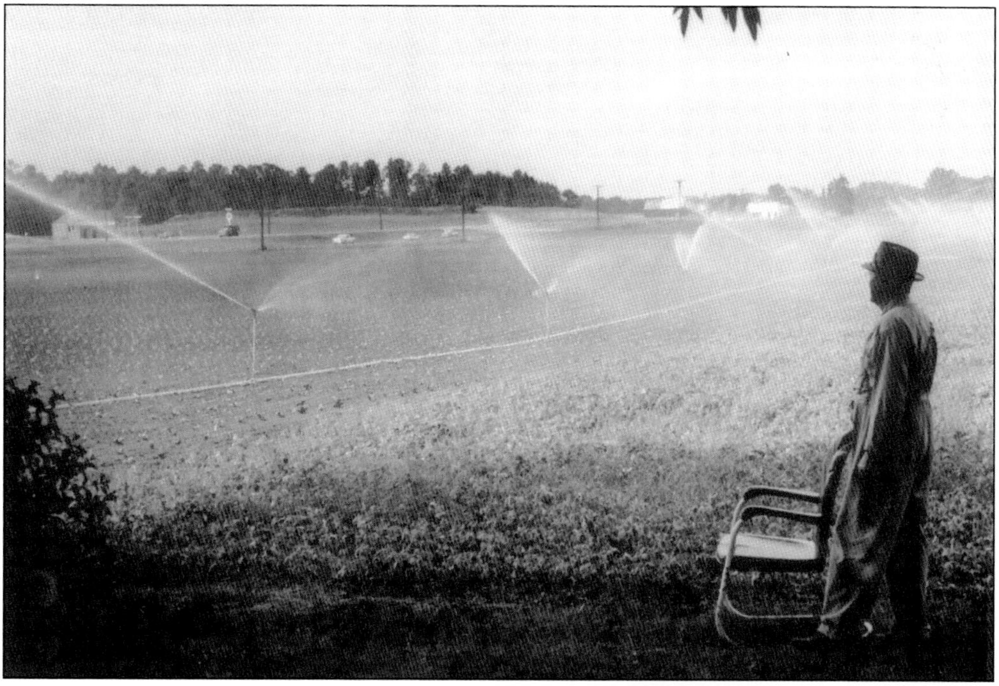

Young tobacco plants quickly get adjusted to their new environment and begin to grow. Farmer Presley S. Taylor Sr. watches his tobacco being irrigated on his farm in Gambrills at Route 3 and Waugh Chapel Road in July 1960. Normally, about 6,500 plants are planted per acre. The plants are spaced 18 inches apart and are arranged in rows that are 42 inches from one another. (Hearst.)

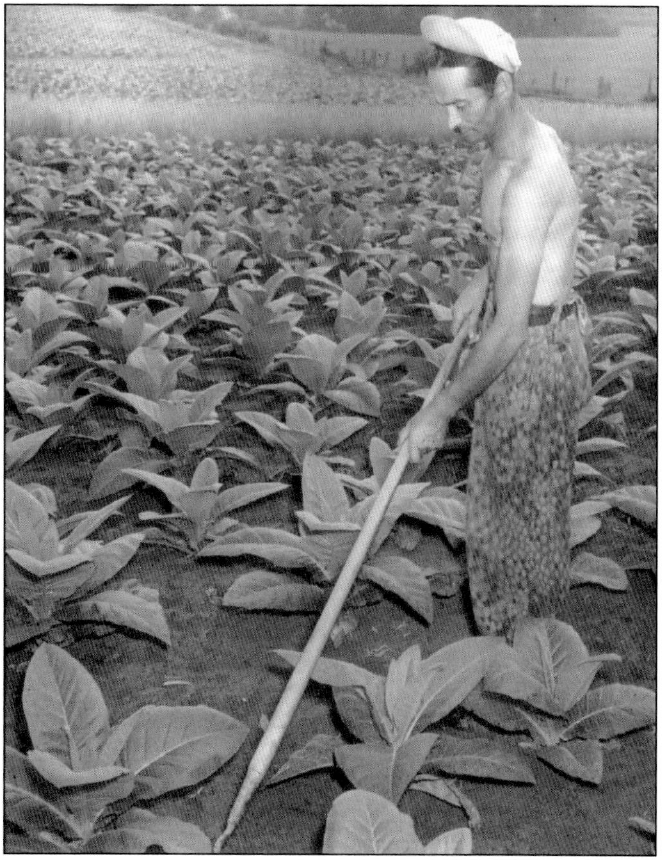

Andre Corrain, a displaced person from the Ukraine, hoes tobacco on the farm of Charles B. Townshend on Davidsonville Road in Davidsonville in July 1949. The tobacco is too large to cultivate by machine, so weeding is done by the use of a handheld hoe. Before mechanization, this was the method used to keep the fields weed-free. (Sun.)

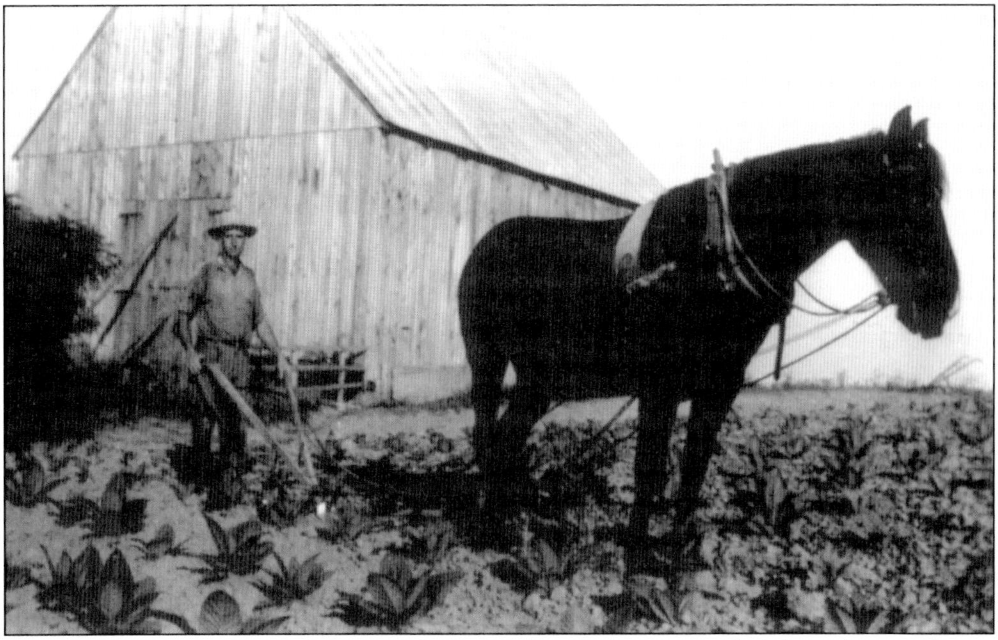

Ernest Rosenauer cultivates his crop of young tobacco plants in early July with a single-row cultivator pulled by a horse on his farm in Crownsville around 1930. His father, Mathias Rosenauer, emigrated from the Banat region of Austria-Hungary in 1906. (Mueller.)

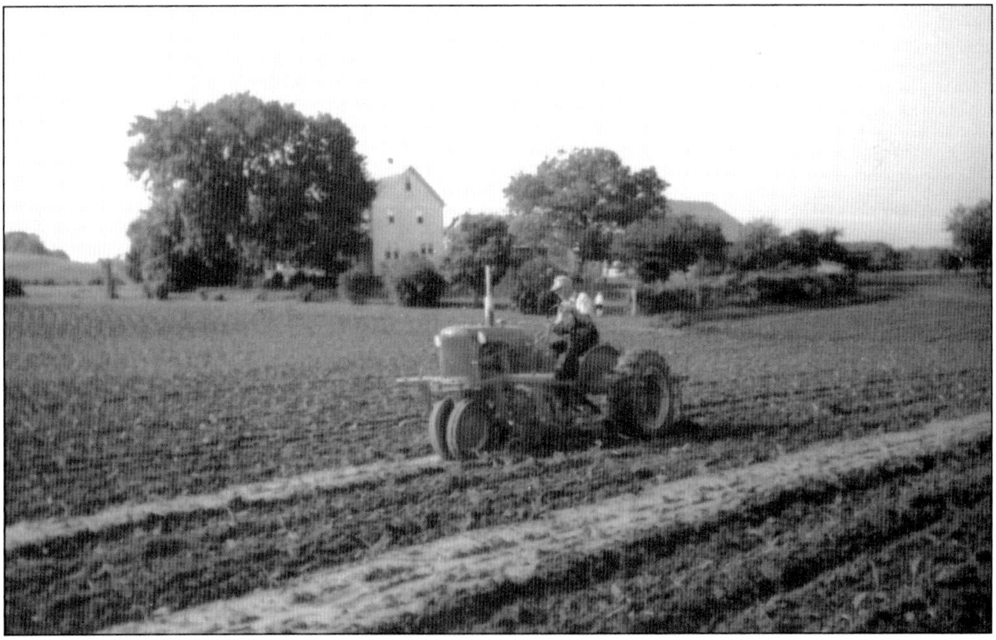

In order to control weeds, farmer William P. Doepkens cultivates his tobacco crop with a two-row cultivator attached to his Allis Chalmers Model C tractor on his Gambrills farm in June 1975. Once planted, tobacco would be cultivated several times before the middle of July to ensure a weed-free field. (FHD.)

Quentin Hittle raised several acres of tobacco on his 50-acre farm in Davidsonville on Governor Bridge Road. Here, Hittle is seen on his horse in 1940. He also used to raise chickens and pigs on the family farm. His dad, Joseph Hittle, emigrated from Budapest, Austria-Hungary, in 1903. (Winterson Hittle.)

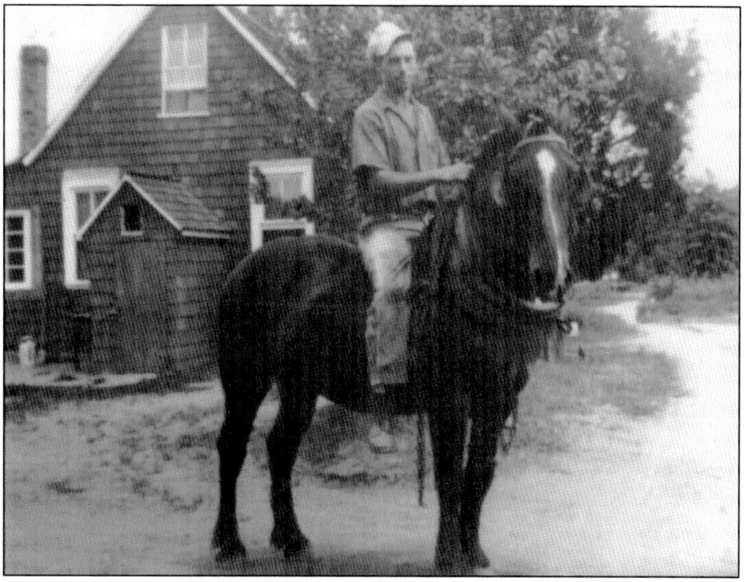

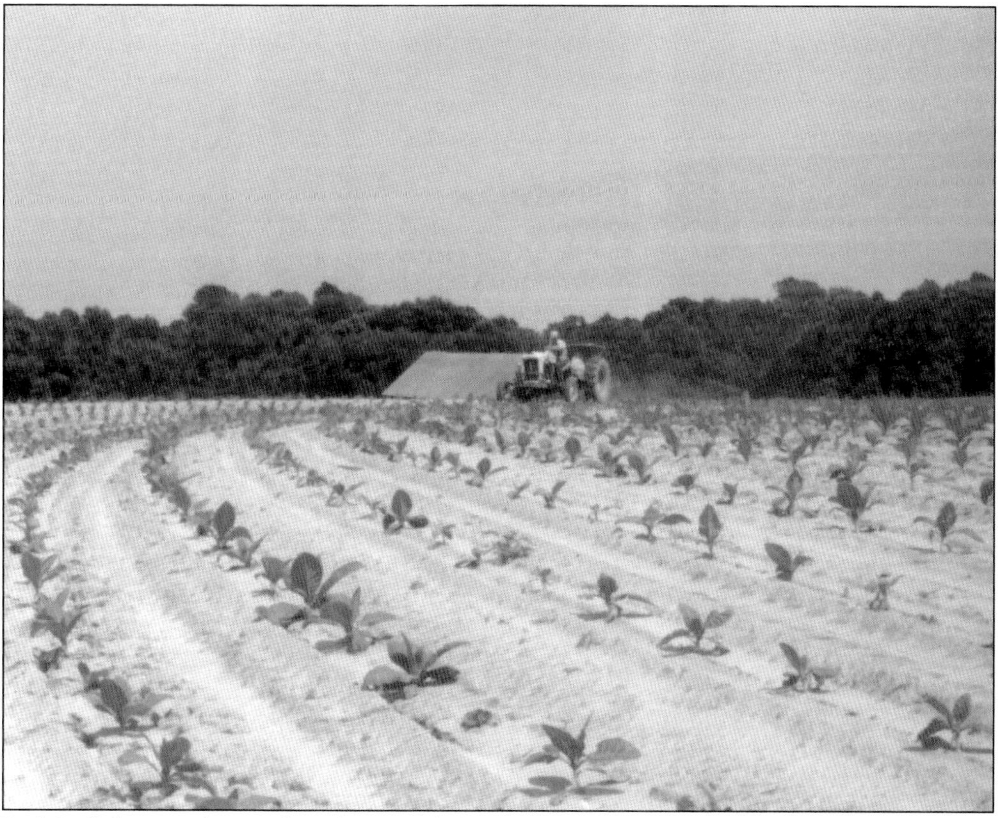

Earl Griffith is top-dressing his tobacco with fertilizer while he is cultivating on his farm in Lothian in July 1983. The Griffiths' farm was the first in the county to install cropland terraces to help control erosion in the fields. In addition to the terraces, the crop rows are planted to follow the contour of the land, which helps prevent soil erosion. (AASCD.)

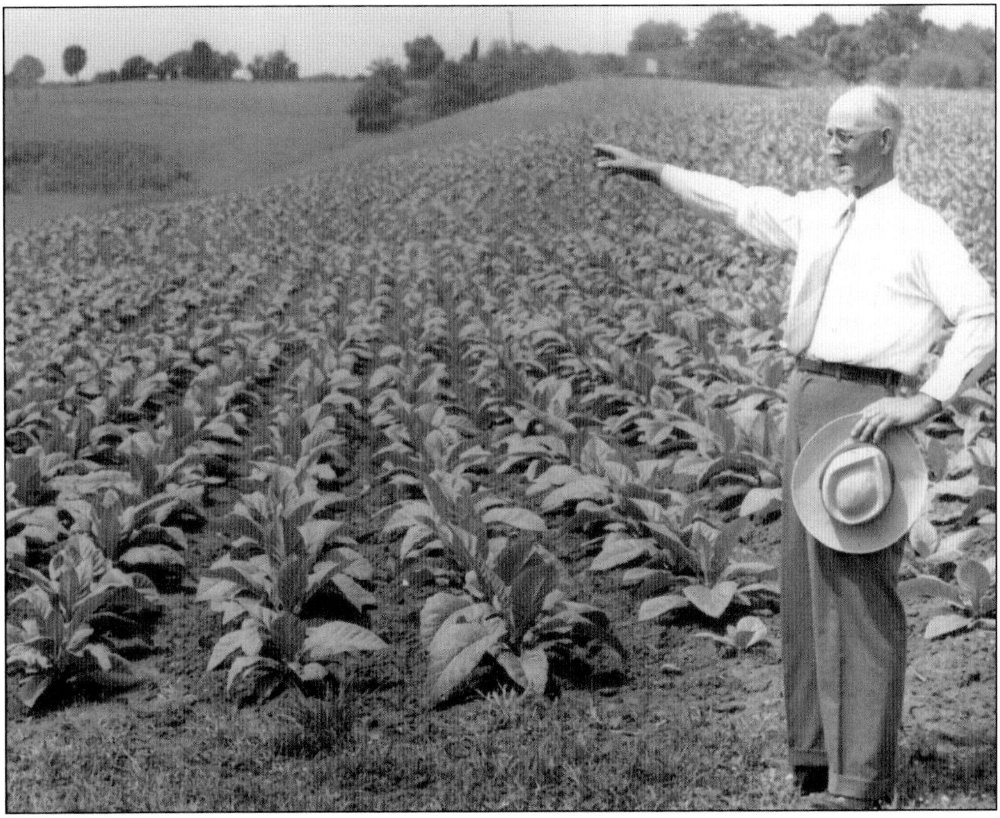

Oscar F. Grimes Sr. inspects his tobacco crop in mid-July 1949 on his farm in Davidsonville. Providing that there is ample rainfall, the plants will grow quickly to maturity in about 70 days. When suitable water is available, some farmers set up an irrigation system in their tobacco fields, which enables them to water their crop during drought conditions to ensure maximum crop growth. (Sun.)

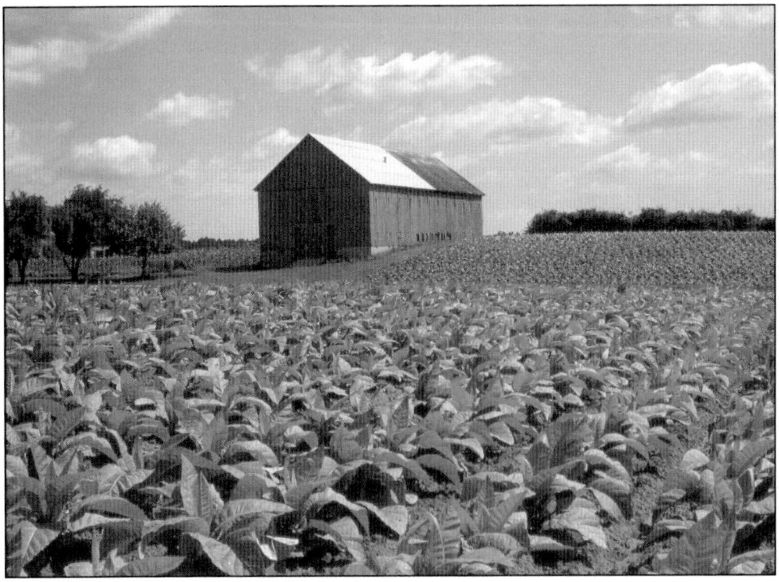

By late July, the plants have grown, and their leaves are covering up the soil, which will help prevent weeds from growing. Pictured in 1980, this field is on the William P. Doepkens farm on Davidsonville Road in Gambrills. Within a couple of weeks, it will be in full bloom. (FHD.)

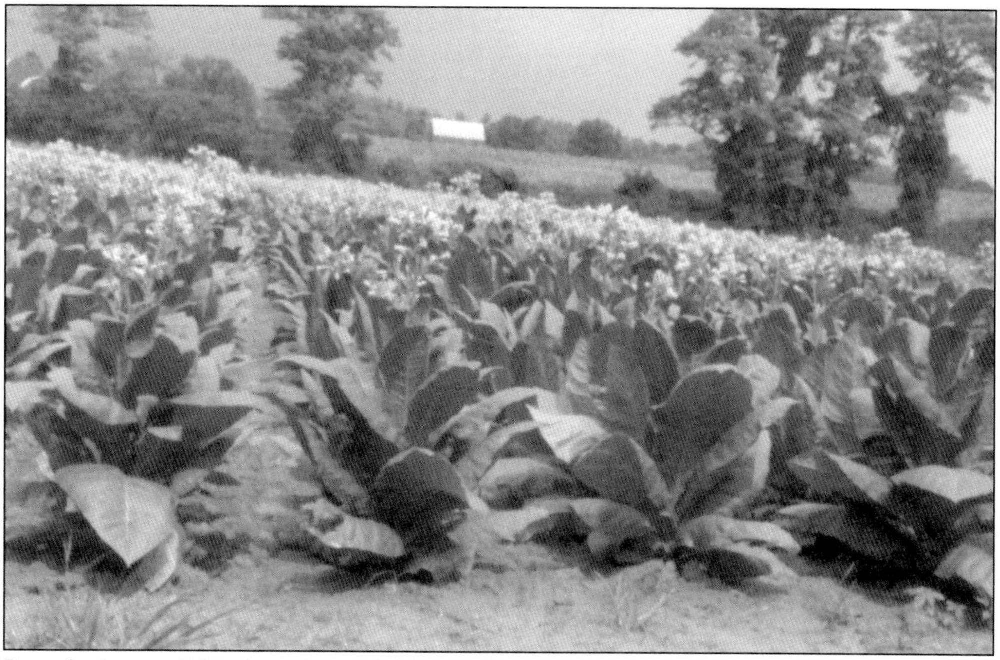

By early August 1956, this tobacco field on Earl Griffith's farm in Lothian is in full bloom. The blooms take energy from the plant, so they need to be removed to allow the lower leaves to grow to a mature size. Sometimes, farmers will keep one or two plants in full bloom and harvest them later for seeds that they can plant the next year. (Griffith.)

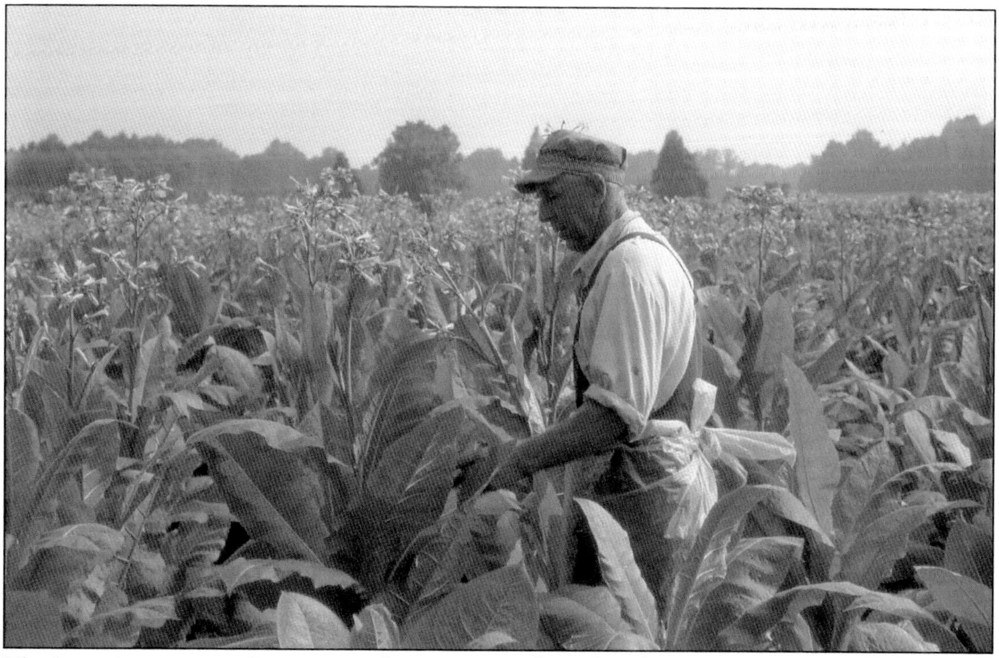

The best time to top tobacco is in the morning, when the plants are still moist and the stems are rigid. The top of the plant (the part with the flower) will be "topped" by hand. Some farmers used a small hooked knife to aid in topping. Pictured in 1979, William P. Doepkens has plastic draped around his waist to keep his clothing dry. (FHD.)

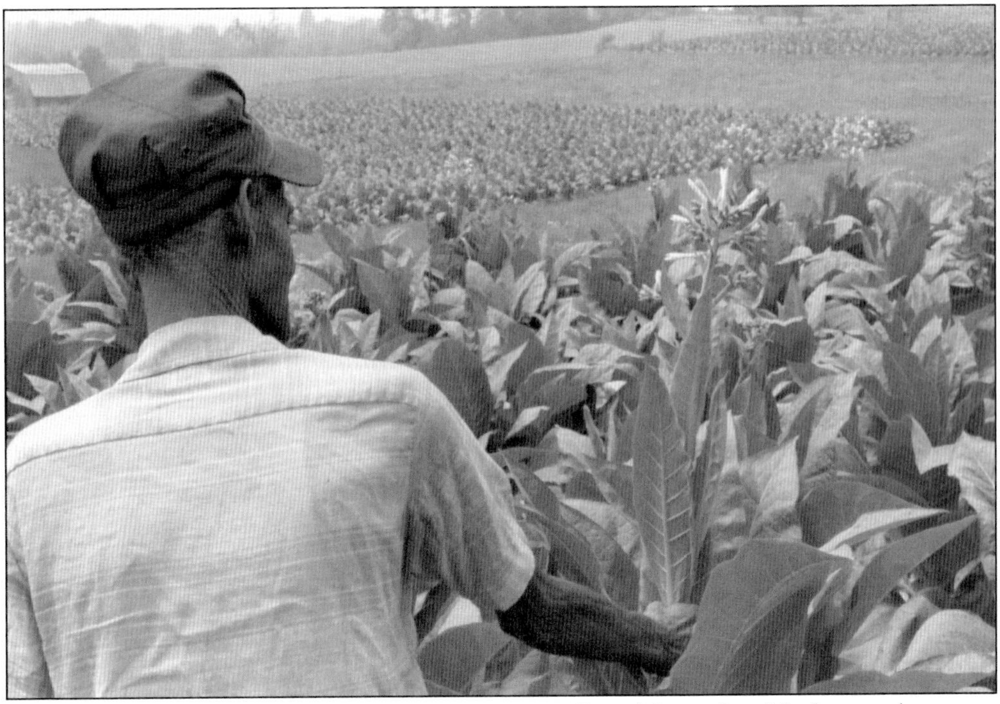

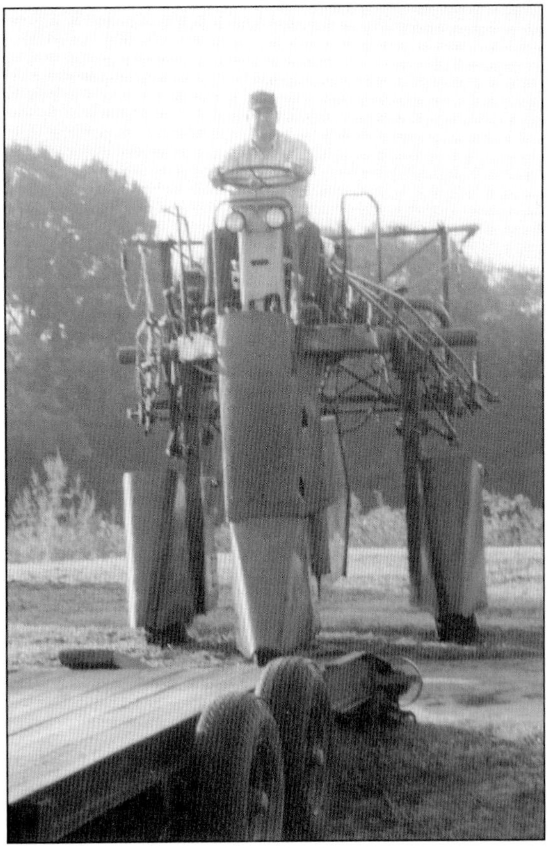

Farm laborer Sam Neal tops tobacco in July 1976 on the William and Laura Zang farm on Muddy Creek Road near Galesville. As he walks between two rows, he will top one row at a time, alternating between the two rows. As the top is removed, it is dropped down onto the ground. Once topped, the tobacco plant will stand three to four feet tall and will have about 18 leaves. (Capital.)

After they are topped, the plants begin to grow suckers. A sucker is a small side shoot that, if allowed to grow, will interfere with the harvesting and curing process. Jeff Griffith, on his farm in Lothian around 1990, is shown on a high bay sprayer, which he uses to spray the tobacco with a growth retardant that will slow down or prevent the growth of suckers. (Griffith.)

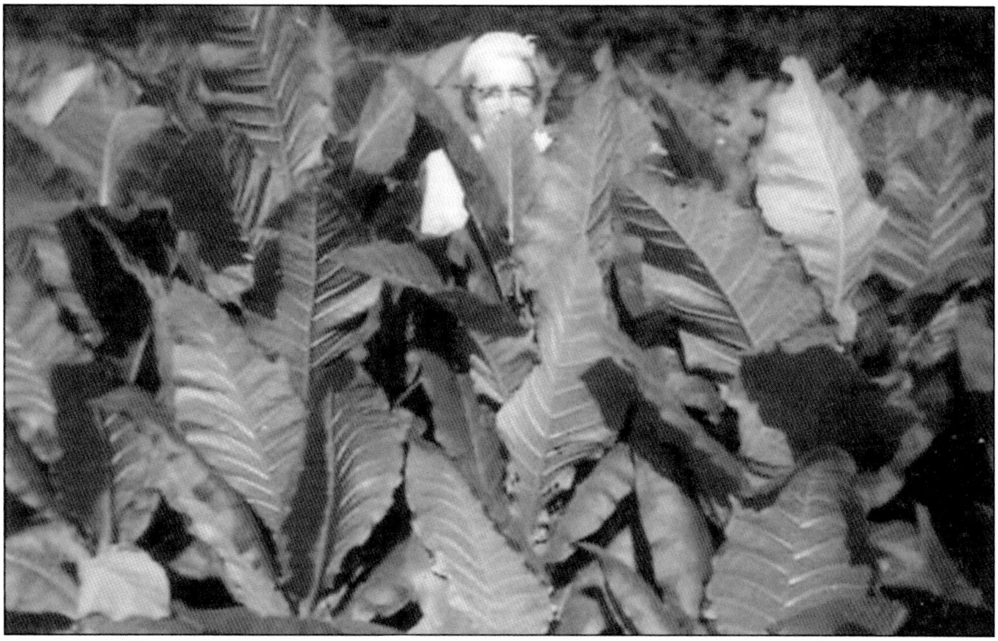

In 1950, Philomena Sommer is standing in her tobacco field, which is ready for harvest. About a week to 10 days after topping, the tobacco is ripened and ready for harvest. John Sommer, her husband, emigrated from Germany in 1914 and bought his 97-acre farm on Davidsonville Road in Gambrills in the 1920s. They raised corn, tobacco, pigs, and free-range chickens. (Helen Sommer Whittington.)

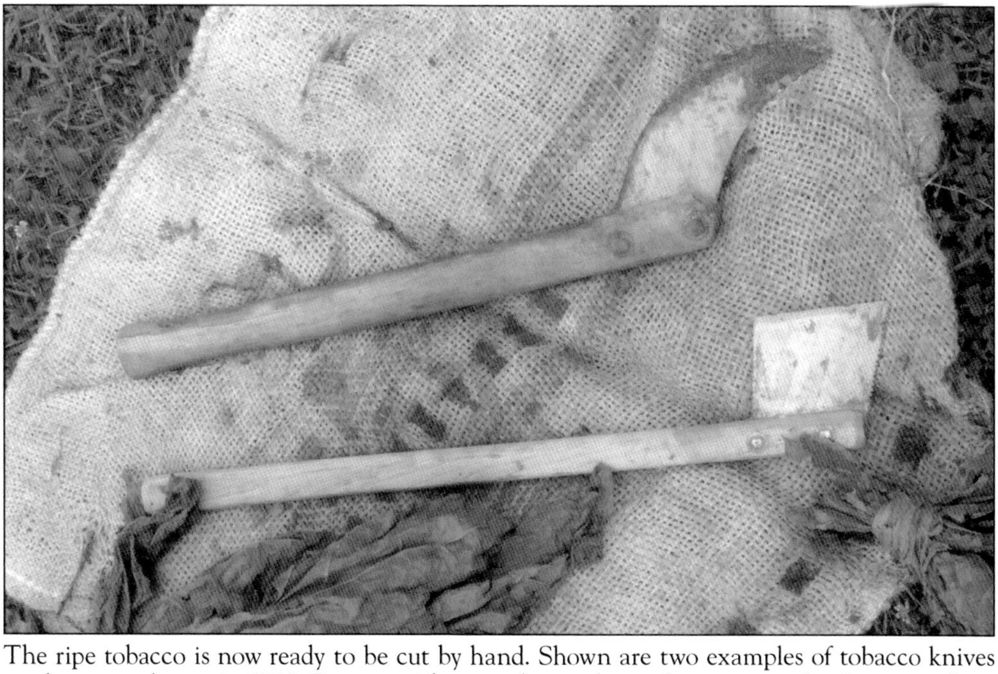

The ripe tobacco is now ready to be cut by hand. Shown are two examples of tobacco knives used to cut tobacco in 2010. Farmers either purchase tobacco knives or make their own from old saws by sharpening a cutting edge on one side. The knives are regularly sharpened using a metal file. (FHD.)

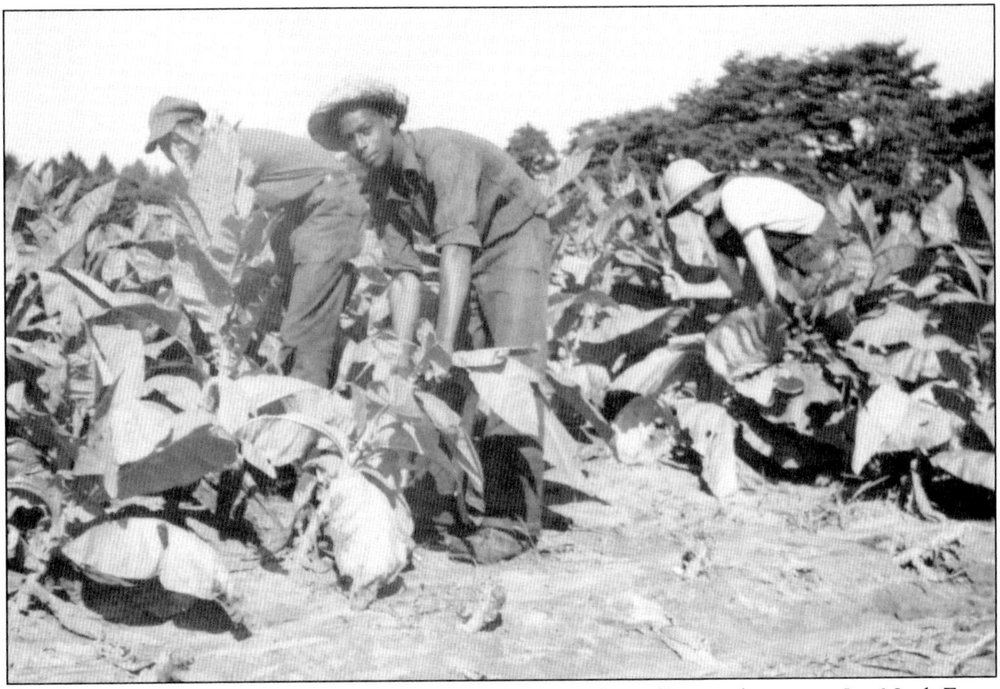

From left to right, August Neal, James Neal, and Mason O'Neill cut tobacco on Ivy Neck Farm around 1940. Two rows are cut at a time. The plants are cut an inch or two above the ground and then carefully placed behind the worker on the ground. Since tobacco is very brittle when wet or cold, farmers will wait until late morning or early afternoon, when the tobacco has dried, to begin cutting. (Colhoun.)

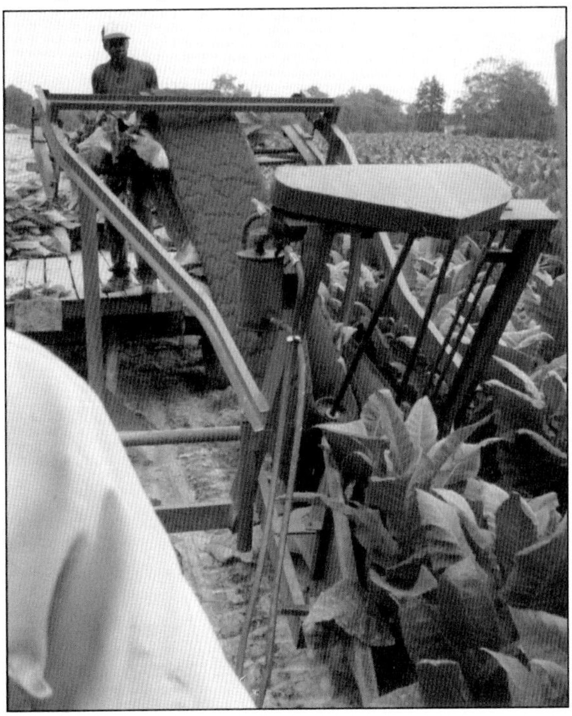

In 1970, the University of Maryland Agricultural Engineering Department developed a tobacco harvester, which was designed to mechanize the tobacco cutting process. The tractor-mounted machine mechanically cut the plants, which were then lifted onto a belt that would place the plants onto a wagon ready for darting. In this picture, the tobacco-cutting machine is being used on the Earl and Jeff Griffith farm in Lothian. (Griffith.)

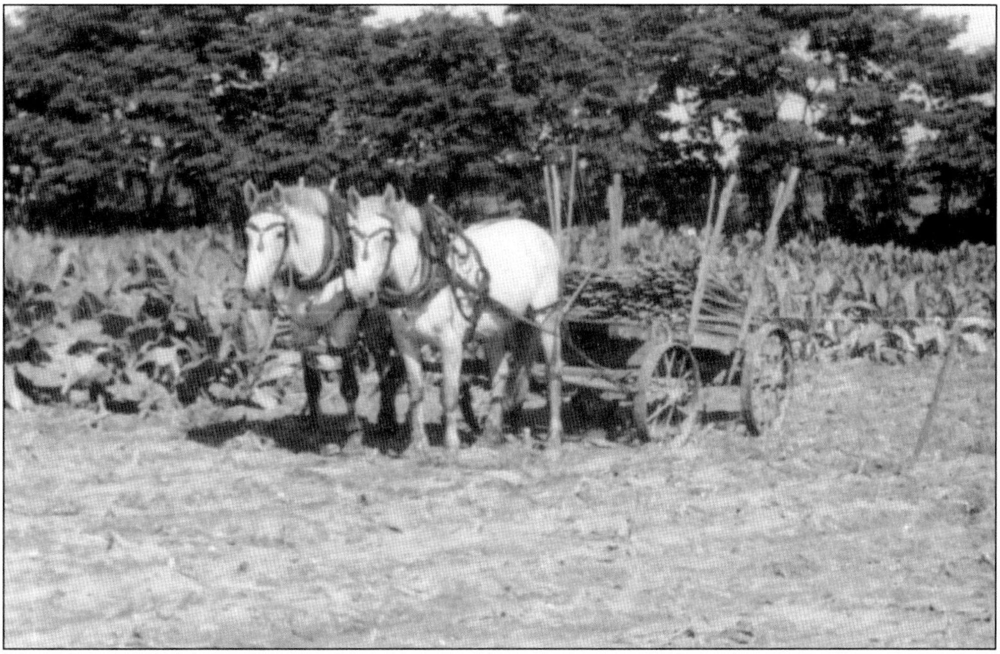

A horse-drawn wagon is loaded with tobacco sticks on the Ivy Neck Farm around 1940. Farmers once made tobacco sticks by hand from oak or poplar wood, but later, the farmer could have them made at a local sawmill. The four- to five-foot tobacco stick would be sharpened on one end. (Colhoun.)

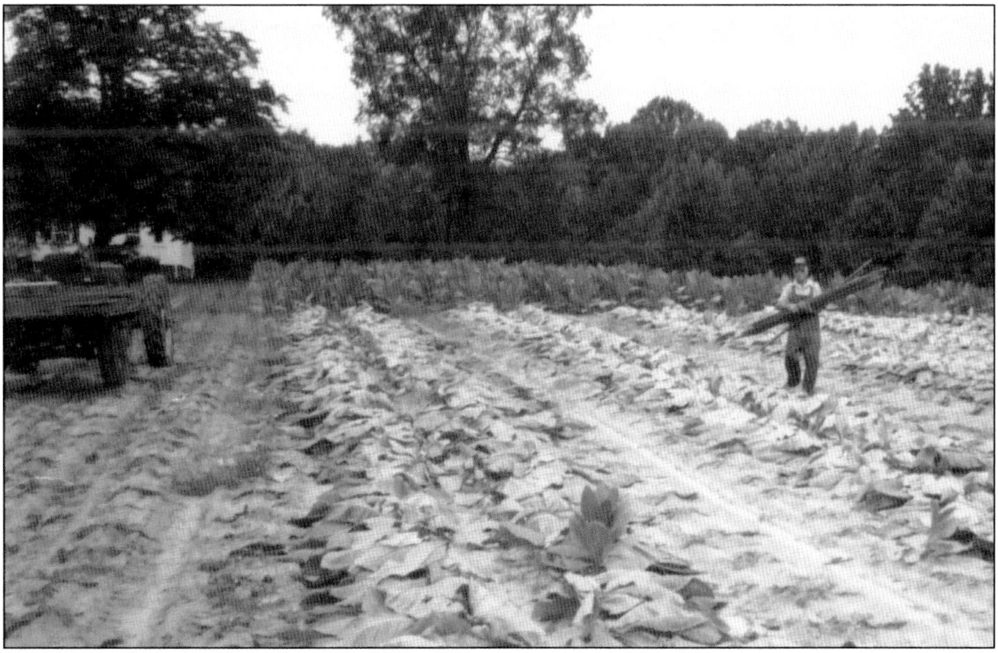

The tobacco sticks are loaded onto a wagon and carried out to the field. Charles Wagner drops sticks in between the rows of the cut tobacco on a rented farm on Governor Bridge Road in 1990. The sticks are overlapped as they are dropped on the ground. His parents, Mathias and Teresia Wagner, emigrated from the Baratzhausen area of Austria-Hungary in 1905. (Peter Wagner.)

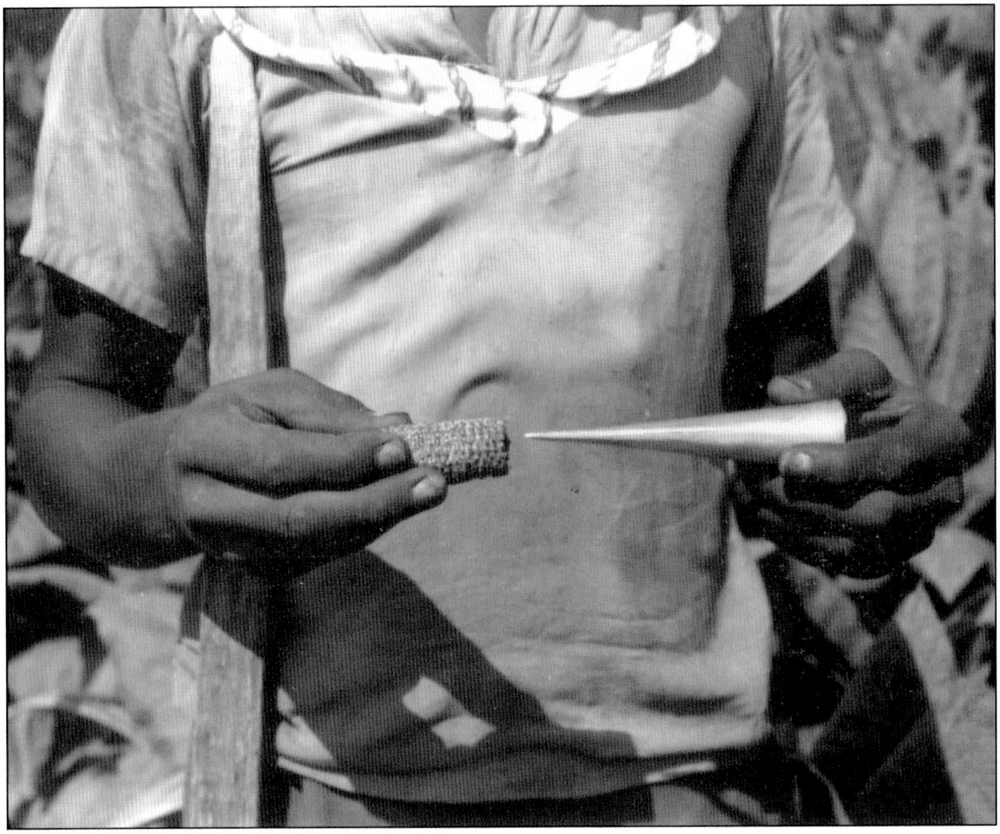

A 1963 tobacco worker displays the tobacco spear or "dart," which is placed on the sharpened end of the tobacco stick. This hand-made or purchased tool allows the tobacco to be easily speared or darted onto the stick. When not in use, a corncob would be placed onto the end of the dart to prevent injury to the worker. (Hearst.)

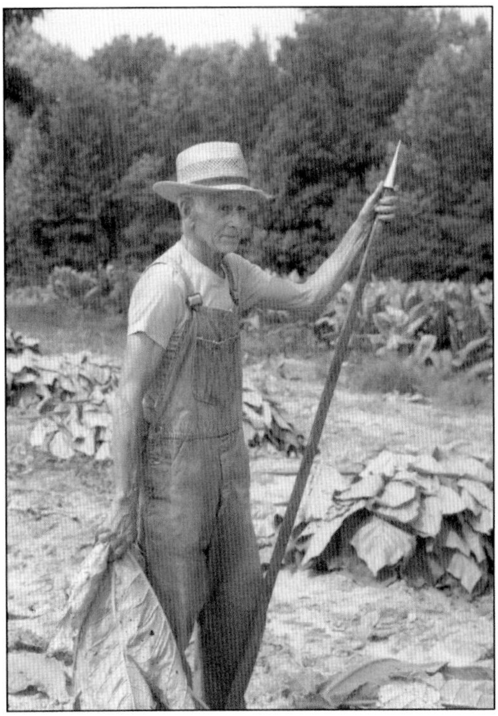

Charles Wagner holds a tobacco stick with a dart on top, ready to spear the tobacco on his farm in Davidsonville in 1990. Depending on size, five or six plants would be speared onto each tobacco stick. If the tobacco is left in the sun for too long, the leaves can get sunburned, which will reduce the quality of the final crop. (Peter Wagner.)

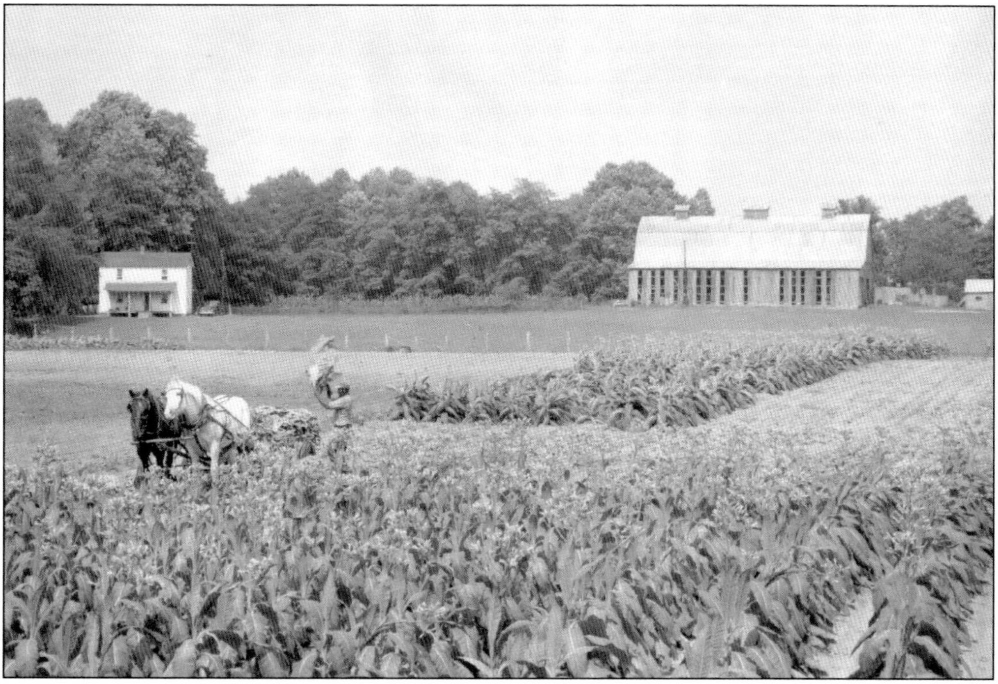

This tobacco farm on Bay Ridge near Annapolis is being harvested by hand and drawn by horses in 1956. In the background one can see the tobacco barn with its slats open for the crop to dry. The community of Chrisland Cove on Bay Ridge Road is now on the site of this farm. (Photograph by Marion E. Warren, © M.E. Warren Photography, LLC.)

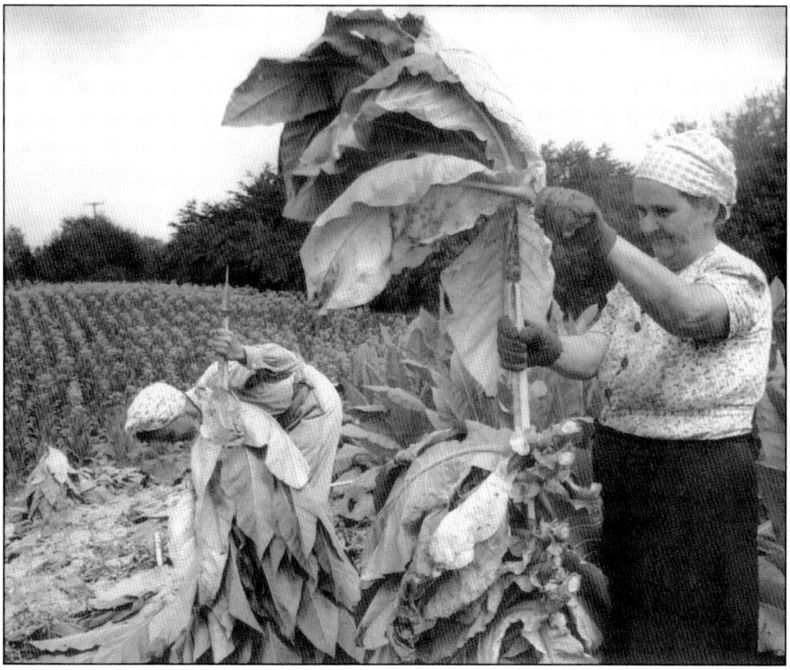

Anne Alberts and her daughter spear tobacco on their family farm in 1947. Anne had two acres under cultivation in Benefield. During World War II, there was a lack of men to work as laborers in the fields, so many women had to help harvest the crops. (Sun.)

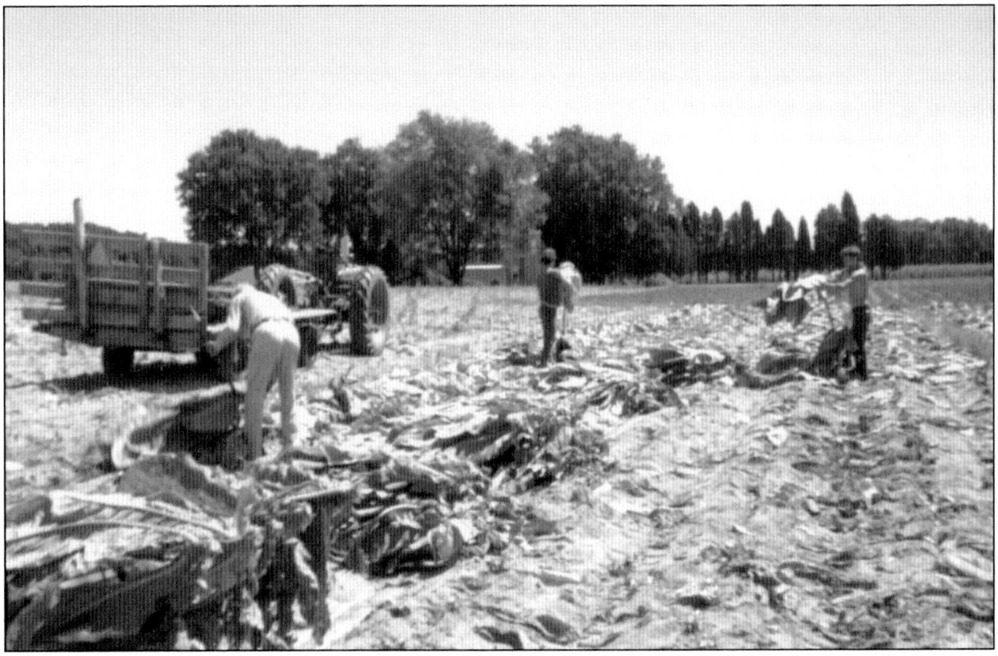

On the Charles D. Pettebone Sr. farm on the Broadneck Peninsula around 1970, he and his son, Charles Pettebone Jr., dart tobacco and then place the sticks of tobacco onto piles, which will allow them to wilt down. The piles are sometimes left overnight and then hung in the barns the next morning. (Roxanne Pettebone McKagan.)

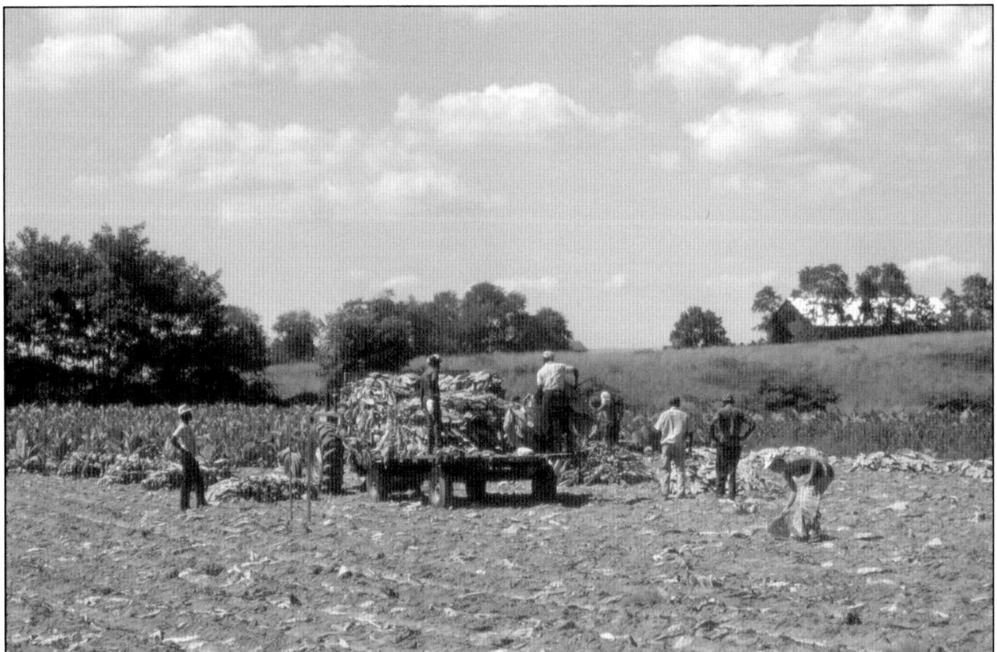

Farm workers load the tobacco piles onto the wagon on the William P. Doepkens farm in 1981. His daughter, Liz Doepkens, is seen behind the wagon picking up leaves that have dropped off the plants. These leaves are valuable and will be tied up, hung in the barn, and cured. (Salyers.)

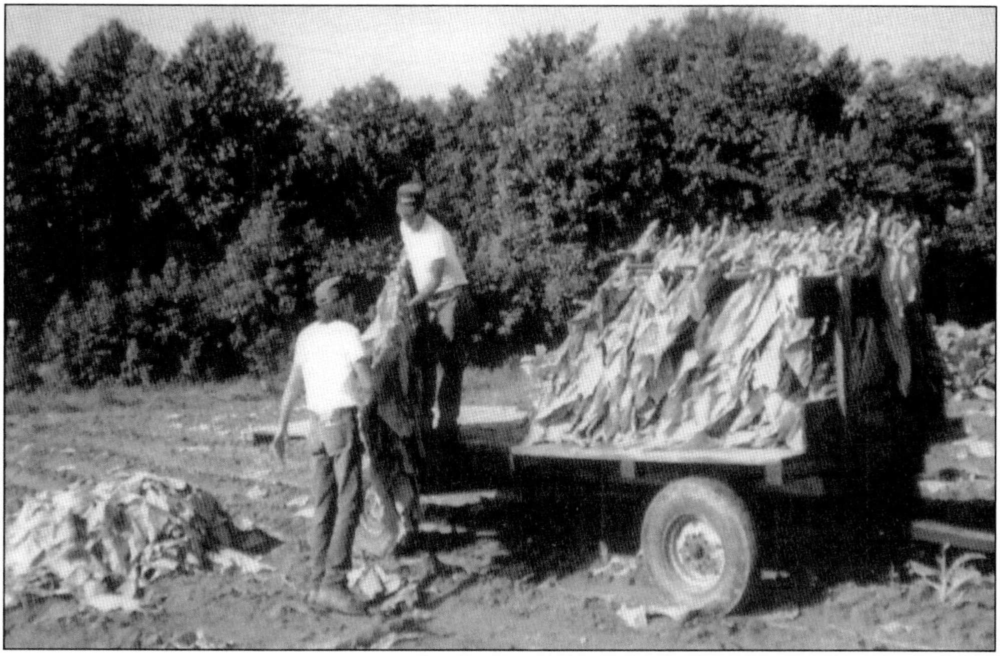

There are different methods farmers use to load tobacco on the wagons in order to carry it to the barn. Eldridge Wilkerson Jr. (on wagon) and Michael Woodridge stack the tobacco in a vertical position on the Eldridge Wilkerson Sr. farm in Harwood in 1992. (Gail and Eldridge Wilkerson.)

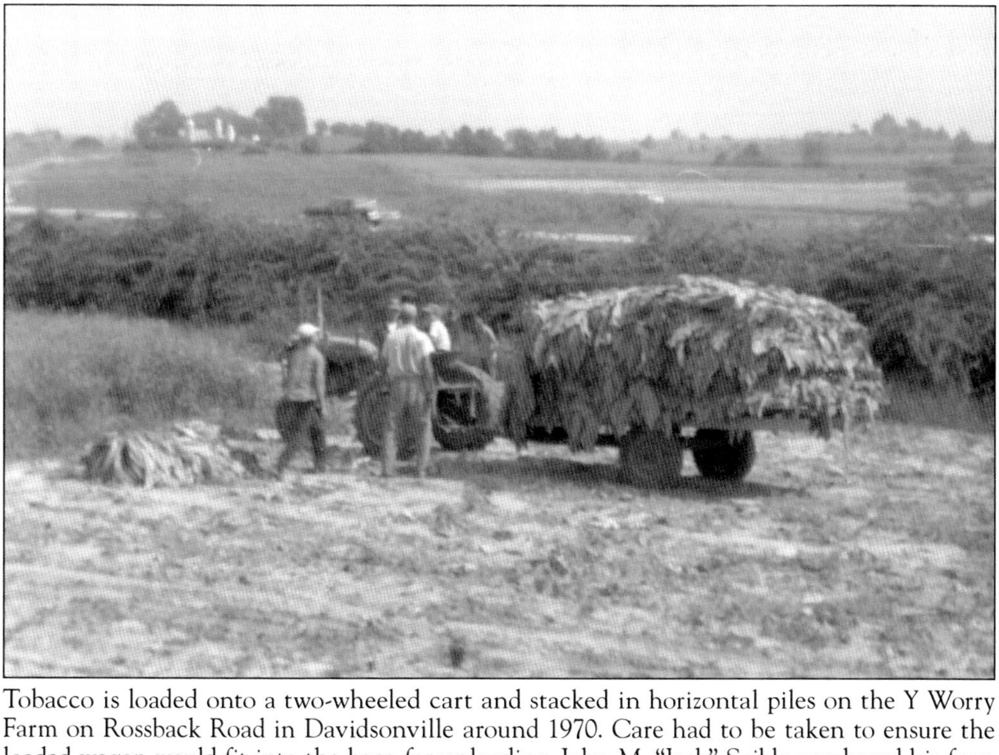

Tobacco is loaded onto a two-wheeled cart and stacked in horizontal piles on the Y Worry Farm on Rossback Road in Davidsonville around 1970. Care had to be taken to ensure the loaded wagon would fit into the barn for unloading. John M. "Jack" Scible purchased his farm in 1953. (Scible.)

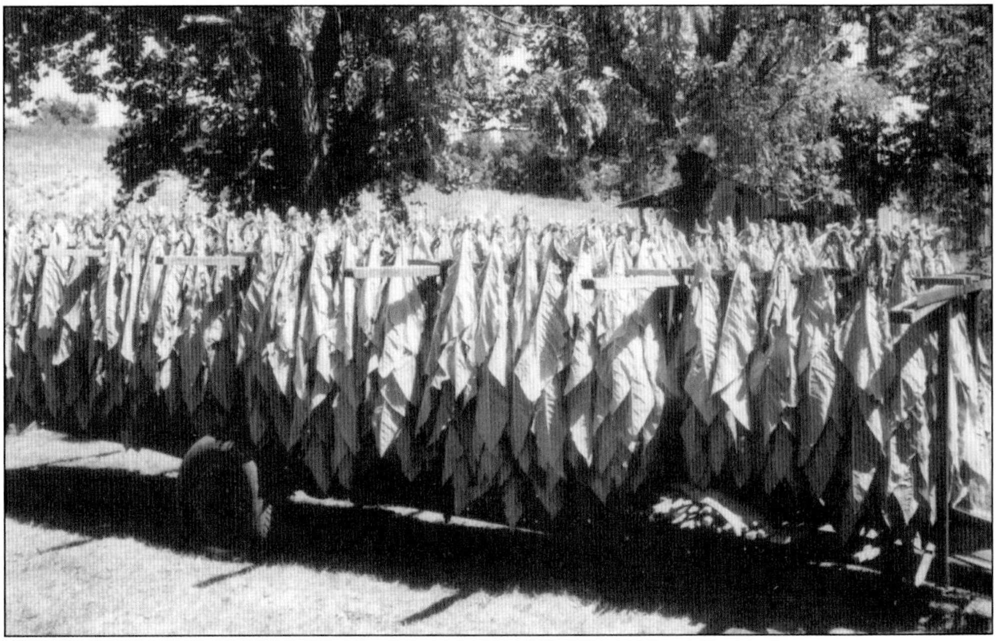

William H. Gertz uses a different method to haul tobacco to the barn on his farm in Crownsville, Maryland, around 1980. The tobacco is hung on racks on a specially made tobacco wagon, which helps to reduce damage to the leaves. Once loaded, the wagon is driven to the barn for unloading and hanging. (Gertz.)

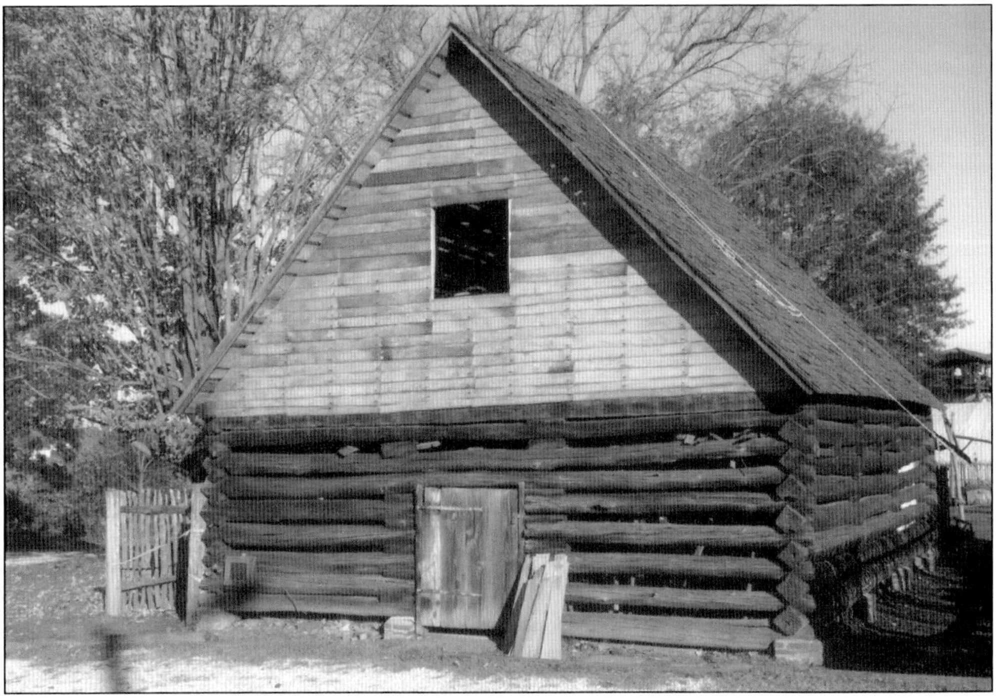

This is an example of a colonial tobacco barn, built in 1720 on the Hockley-in-the-Hole Plantation near Annapolis. It is now on the property of the Historic London Town and Gardens in Edgewater. The space between the logs provided ventilation needed to air-cure the tobacco properly. (FHD.)

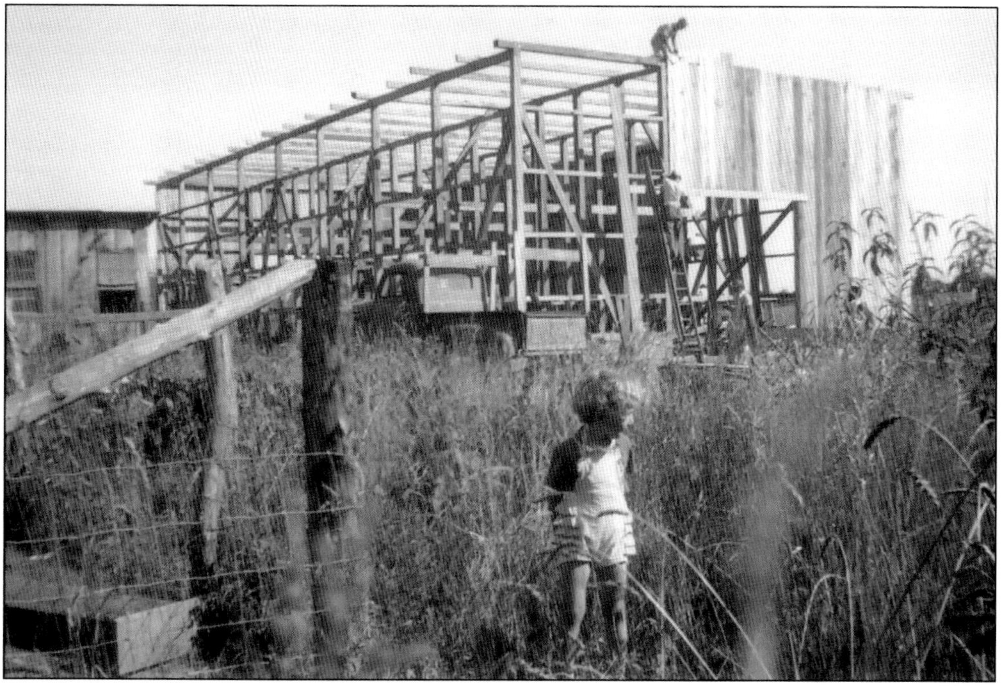

As new tobacco barns were needed, farmers either built them or paid to have them built by local carpenters. Mary T. Scible watches as a new tobacco barn is being built by Nick Bottner on the Henry Doepkens farm in Gambrills using lumber harvested from trees on the farm around 1952. Nick and George Bottner were local carpenters who built many barns and farm buildings in the Davidsonville area. (Scible.)

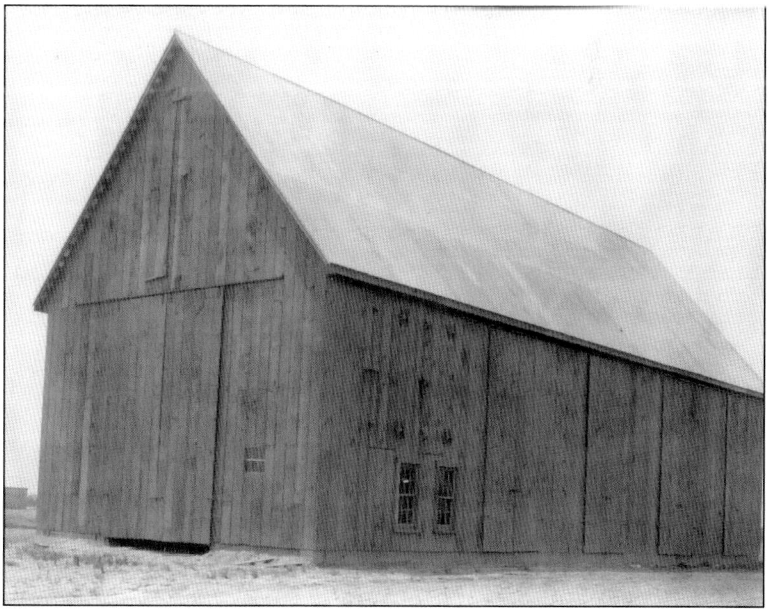

This is an exterior view of a typical southern Maryland tobacco barn. This barn has a gable roof, hinged side vents, and a door on each gable end of the barn. The windows allow natural light into the stripping-room area of the barn. The outside of the barn is covered with wooden boards and usually has a metal roof. (UMCP.)

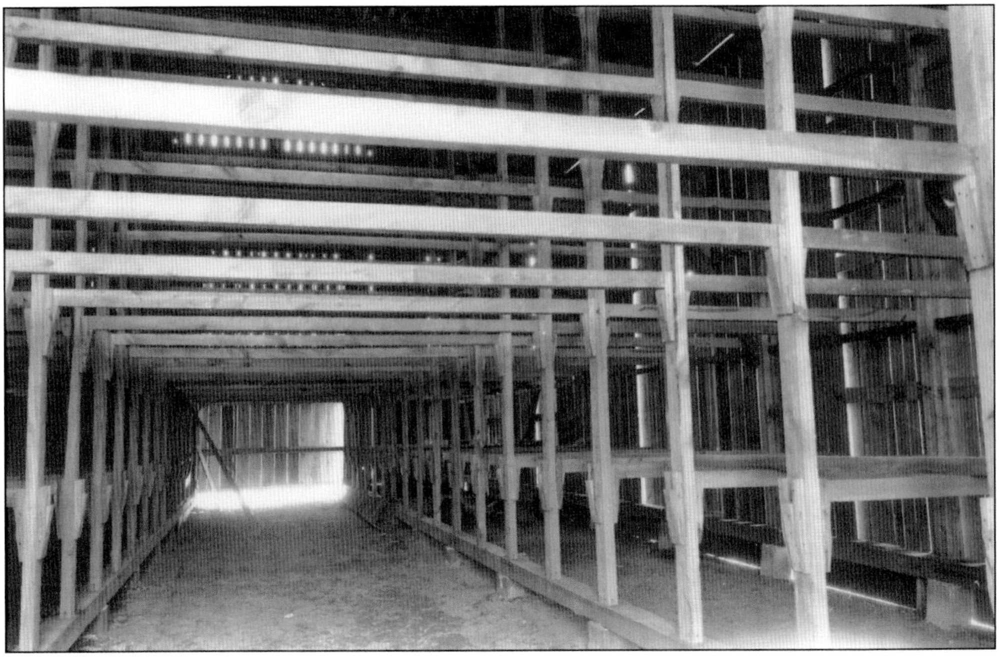

This is an interior view of a Maryland tobacco barn. The barn consists of rooms spaced about four or five feet apart. Each room is divided into tiers as one moves higher up the barn. The bottom poles for the tiers can be removed to allow a load of tobacco to be driven into the barn. The barn usually has a dirt floor. (UMCP.)

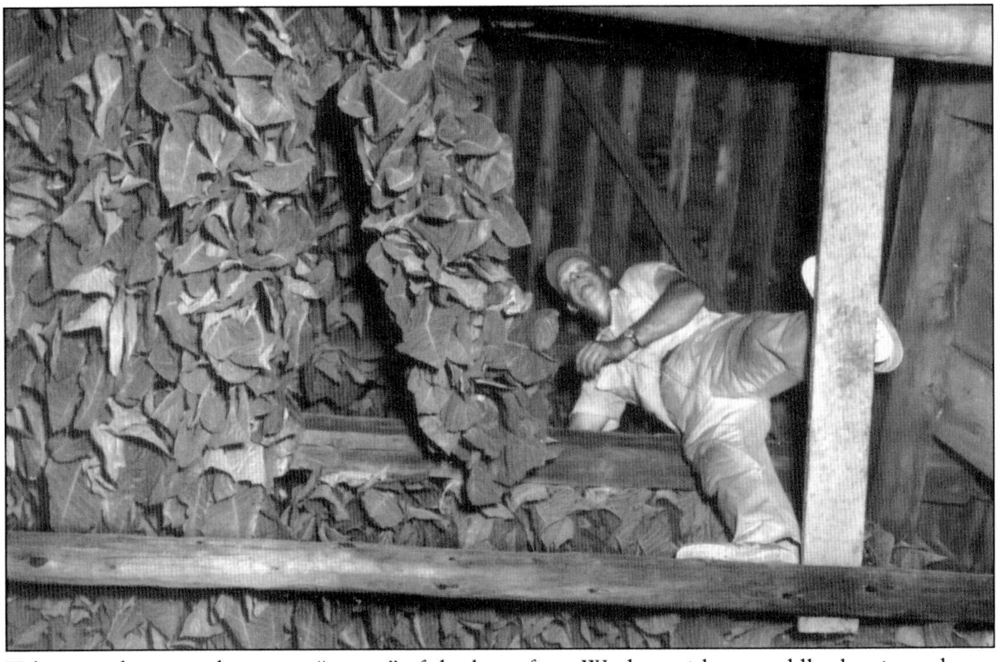

Tobacco is hung in the top or "crown" of the barn first. Workers either straddle the tier poles or stand on boards at various levels in the barn, as pictured here in 1965. Each stick of tobacco is handed up from the wagon to a person on each tier until it reaches the desired location in the barn, where it is hung on the poles. (Sun.)

A tobacco barn on the Jeff and Earl Griffith farm is being crowned out around 1980. Four men are needed in the barn as the loaded sticks of tobacco are handed up to workers in each tier, where they are hung across the poles. The sticks are spaced nine to twelve inches apart, which will permit proper air circulation for curing. (Griffith.)

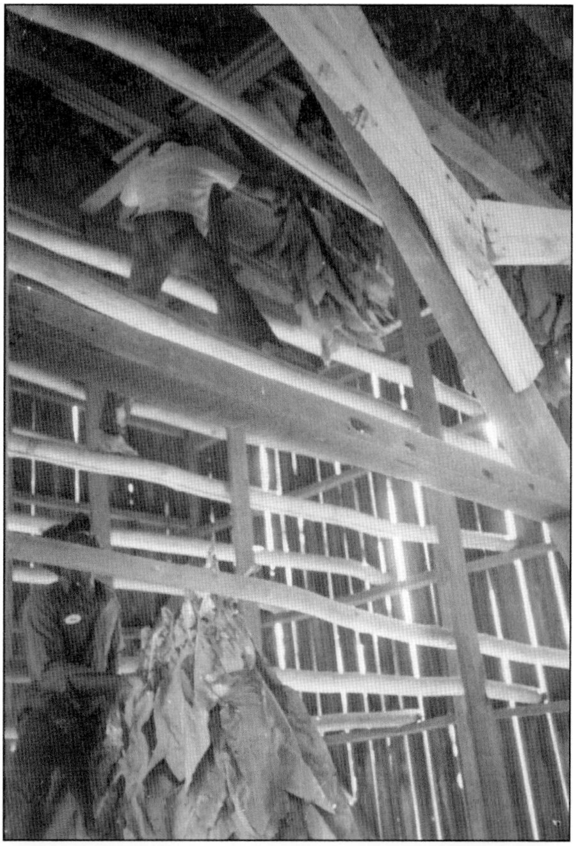

This is an example of a fully hung tier in a tobacco barn on Governor Bridge Road in Davidsonville on the Fricke farm, which was farmed by Charles Wagner and his son Peter Wagner in 1990. The side doors are opened to help provide light and ventilation while the tobacco is being hung. (Peter Wagner.)

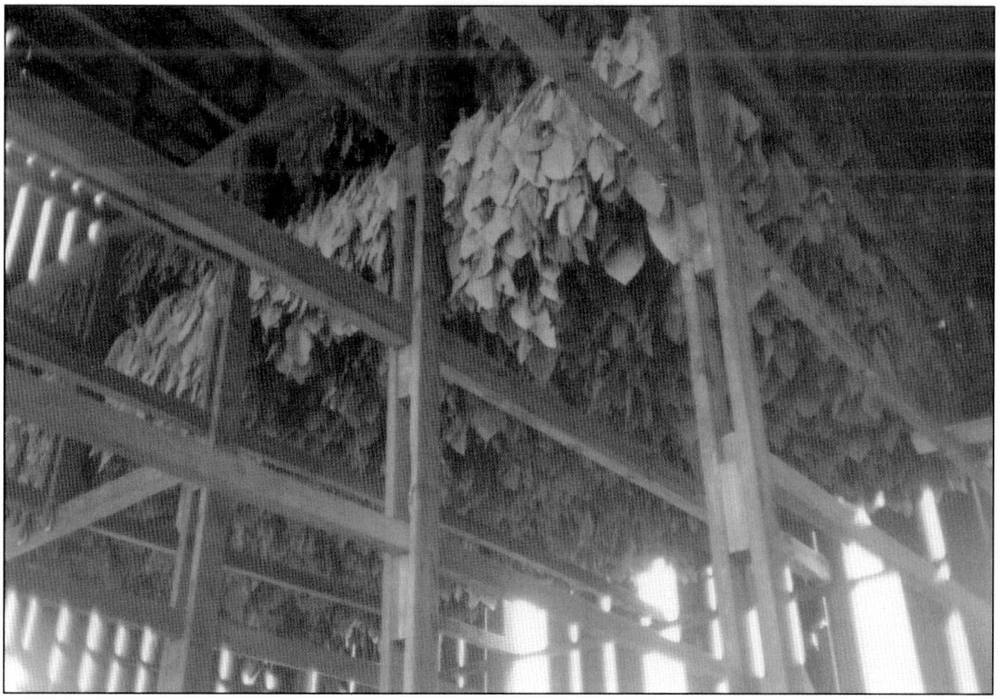

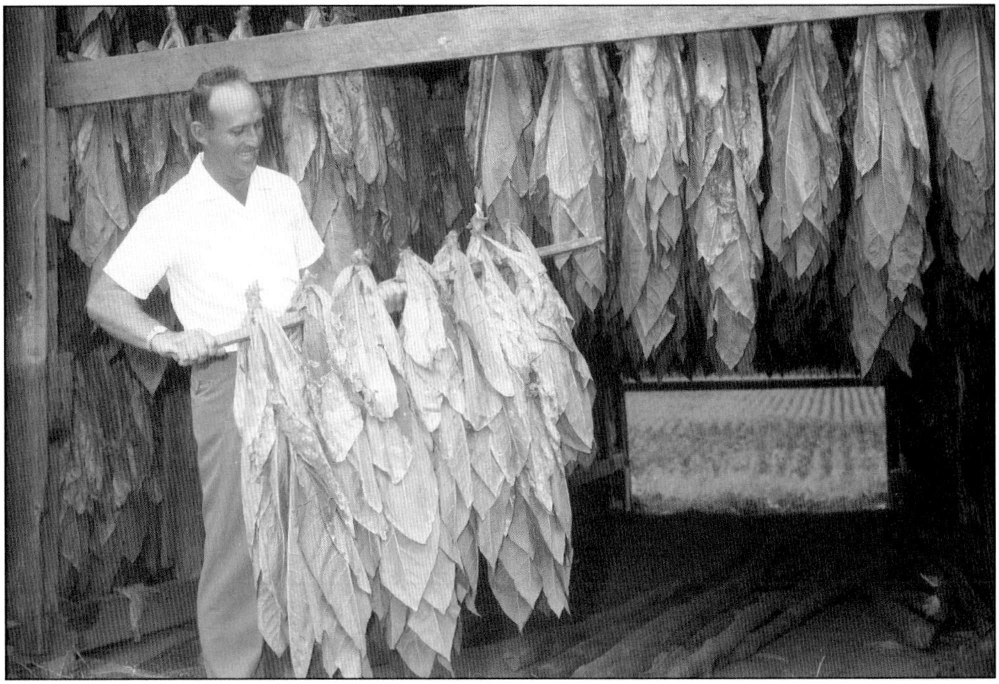

Bernard A. Doepkens Sr. holds a stick of tobacco in front of his barn on his farm in Harwood around 1965. The number of tobacco stalks placed on the stick depends on the size of the plants. In addition to his farm in Harwood, he also worked as a salesman for the Maryland Tobacco Growers' Association. (AACHS.)

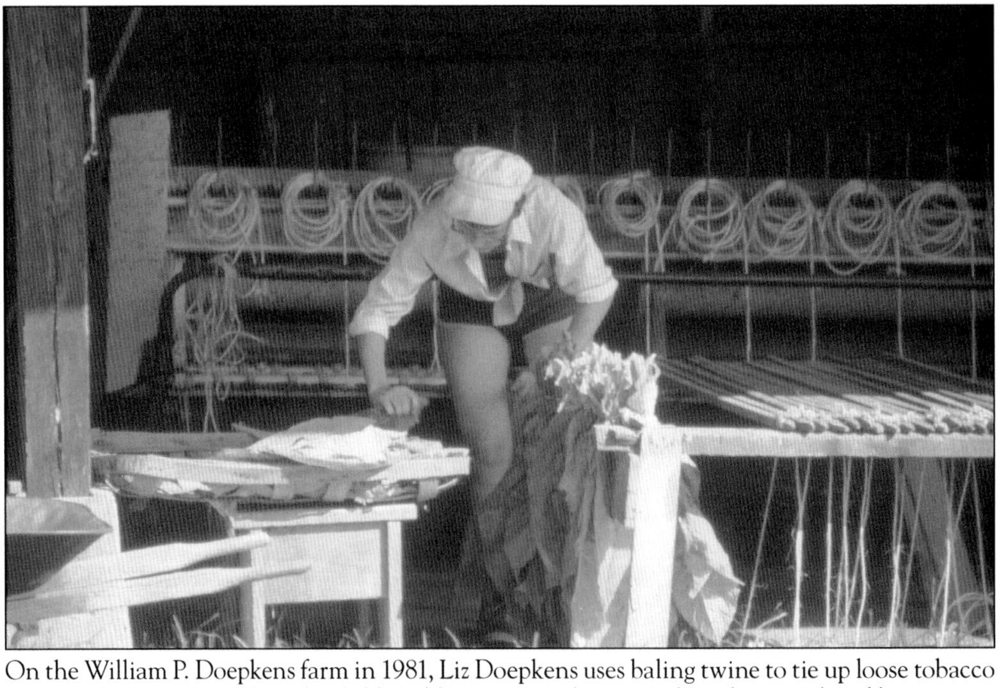

On the William P. Doepkens farm in 1981, Liz Doepkens uses baling twine to tie up loose tobacco leaves that she collected in the field and barn onto tobacco sticks. These sticks of leaves were then hung in the barn and dried. (Salyers.)

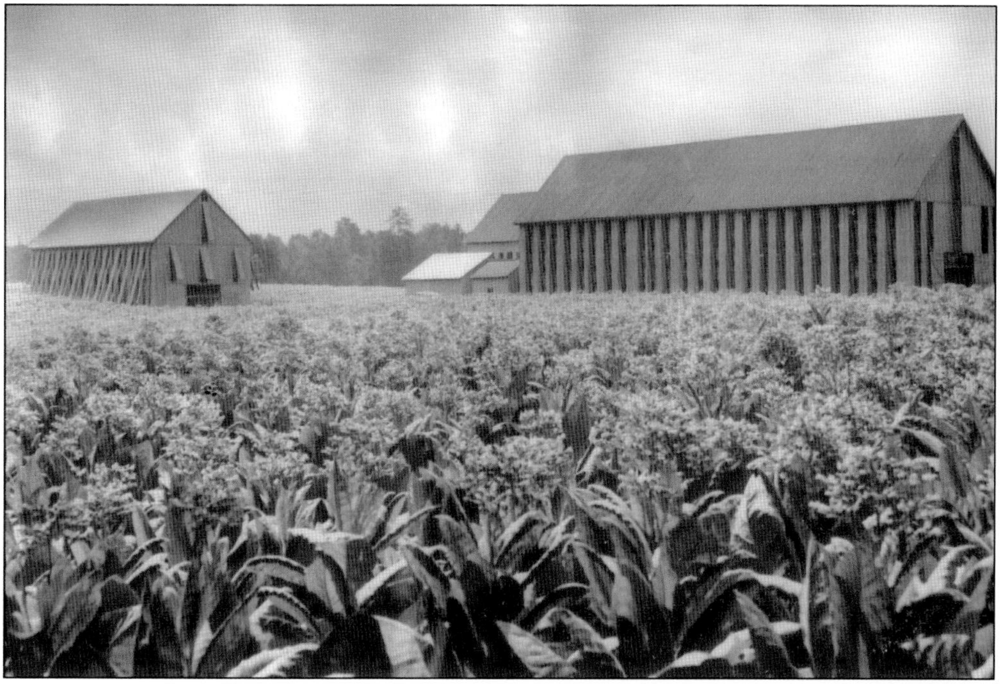

Once the tobacco barn is fully hung, the curing process begins. The sides of the barn are louvered so that they can be opened to allow natural air curing. This system is unique to tobacco grown in Maryland. On the 40-acre Rufus Griffith tobacco farm in Glen Burnie, the barns are fully hung and beginning to dry in 1950. (Sun.)

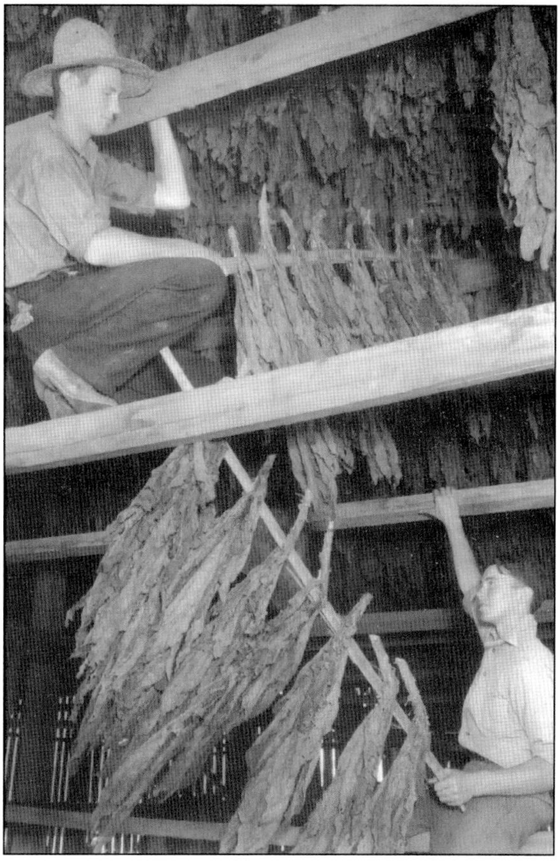

By October 1947, the tobacco is fully cured and ready for stripping. Here, the workers are taking the sticks of dried tobacco down and getting the tobacco ready to go to the stripping room. This can only be done when the tobacco is in order. During rainy or foggy days, the tobacco absorbs moisture and becomes less brittle and therefore can be handled. (Sun.)

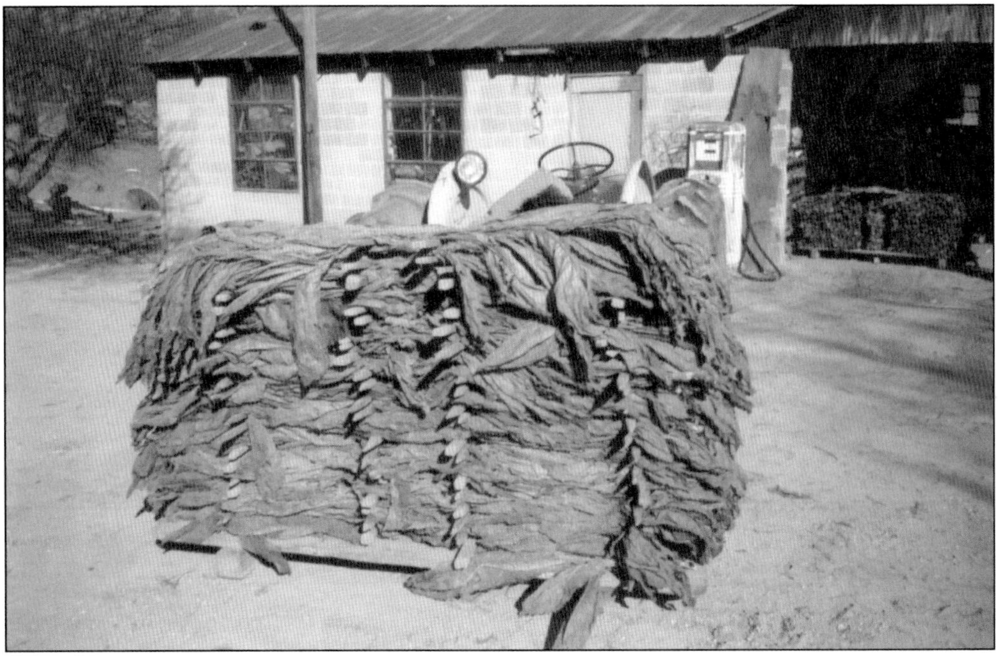

On the William Gertz farm in Crownsville around 1980, the dried sticks of in-order tobacco have been stacked in piles on a pallet to allow easy transport to the stripping room. Once there, the tobacco stalks are removed from the sticks and placed in piles. Each pile is then covered with plastic to keep it in order until it is brought into the stripping room. (Gertz.)

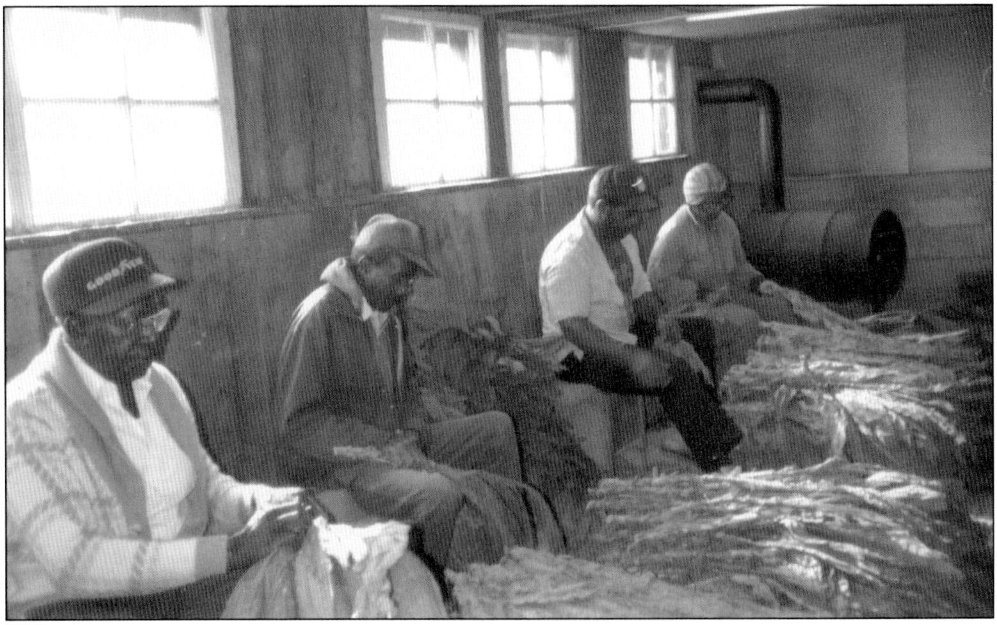

In the stripping room in the tobacco barn on the William P. Doepkens farm, workers sit on benches while they strip tobacco in 1975. Stripping rooms have windows for natural light but may also have artificial light. Since the stripping is done during the winter months, the rooms are heated to help keep the workers comfortable. Stripping accounts for about half the labor for the entire crop. (FHD.)

John Gertz strips the leaves off the stalk one at a time on his farm in Crownsville in 1976. The leaves are graded for quality as they are stripped. There are usually five grades of tobacco: starting at the base, there are ground leaves, and moving up the plant, there are seconds, brights, dulls, and then tips. The best grade is the brights. (Gertz.)

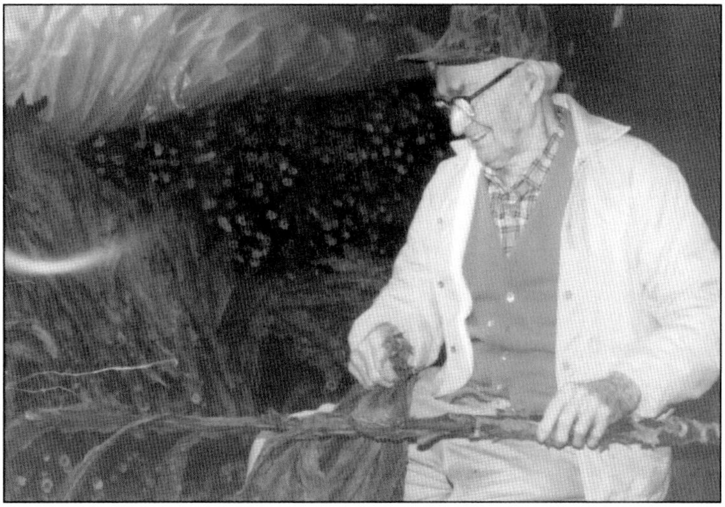

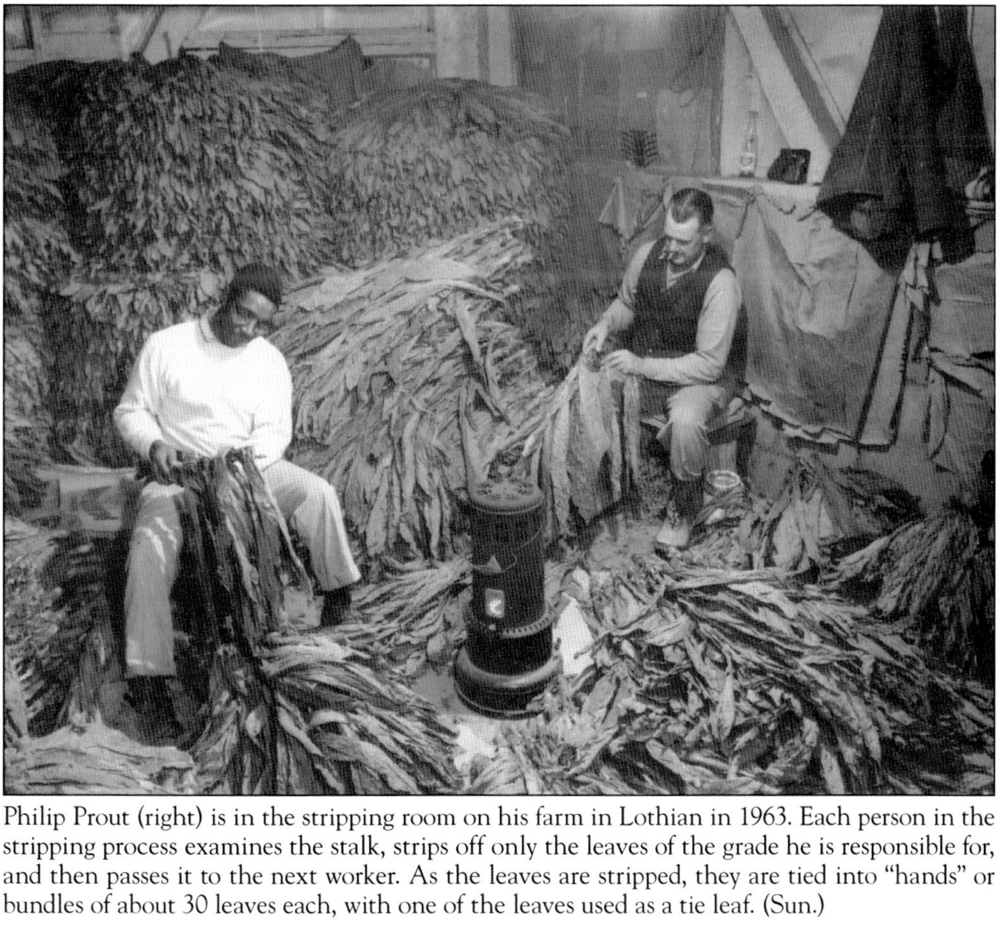

Philip Prout (right) is in the stripping room on his farm in Lothian in 1963. Each person in the stripping process examines the stalk, strips off only the leaves of the grade he is responsible for, and then passes it to the next worker. As the leaves are stripped, they are tied into "hands" or bundles of about 30 leaves each, with one of the leaves used as a tie leaf. (Sun.)

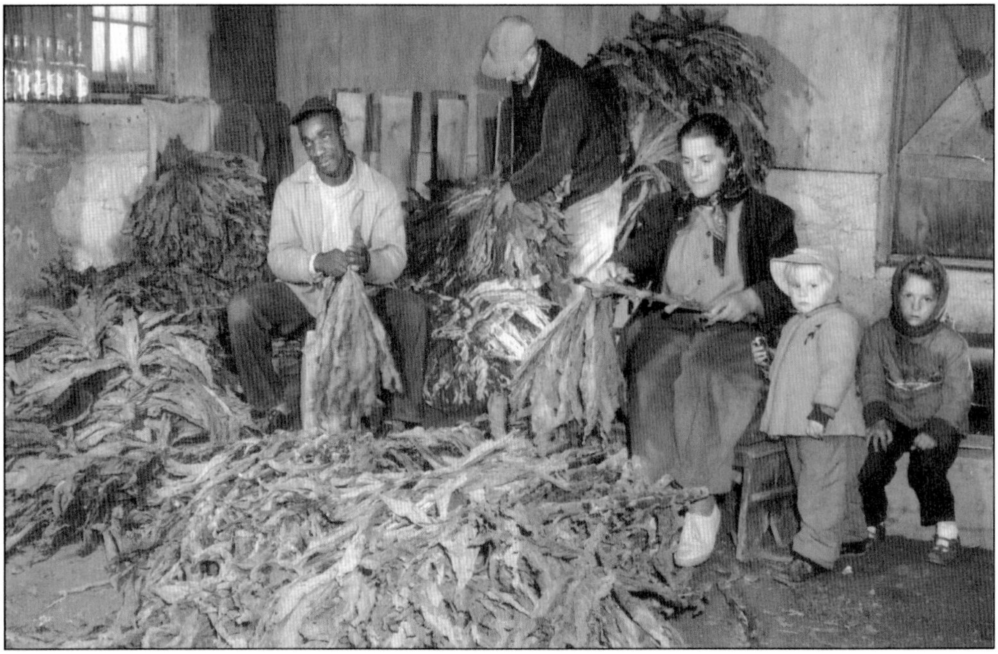

On Benjamin Watkins III's Locust Farm in Davidsonville, the job of stripping tobacco is underway in 1952. Workers, from left to right, are Charles Johnson, Benjamin Watkins III (owner), and Maria Shepet with her children. Mrs. Shepet was a displaced person from the Ukraine working on the farm. (Sun.)

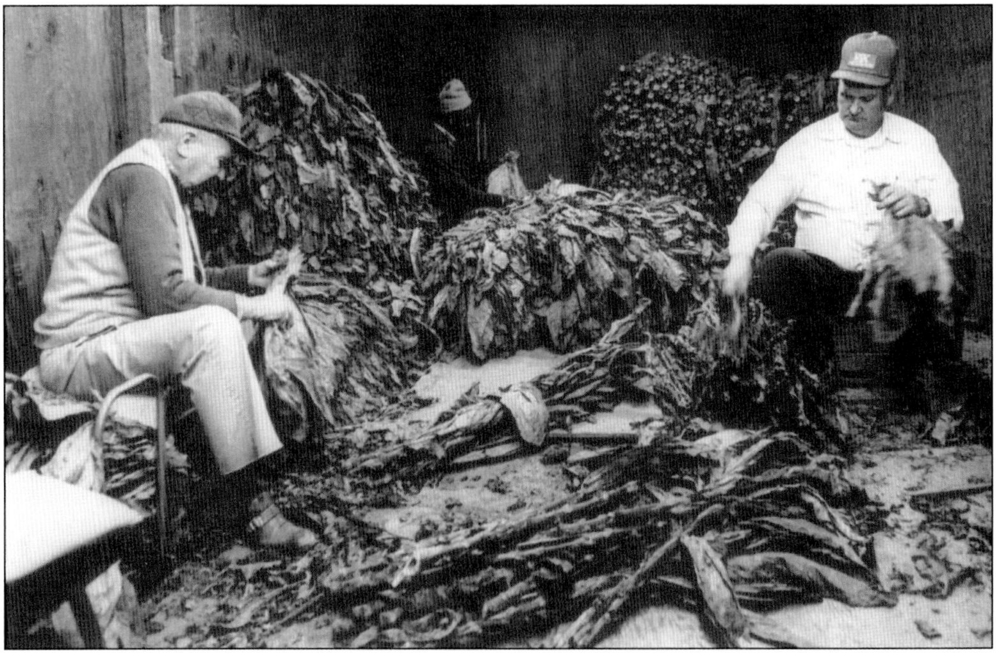

Eugene Griffith (left), Russell Dandridge (center), and Earl Griffith (right) strip tobacco on the Griffith farm in 1985. During the 1980s, a new alternative to tying the leaves into hands was used on farms. Here, Eugene Griffith strips off the leaves from the stalk and places them loose onto a cart called a Handy Pack. These leaves are later packed into bales of tobacco. (Hearst.)

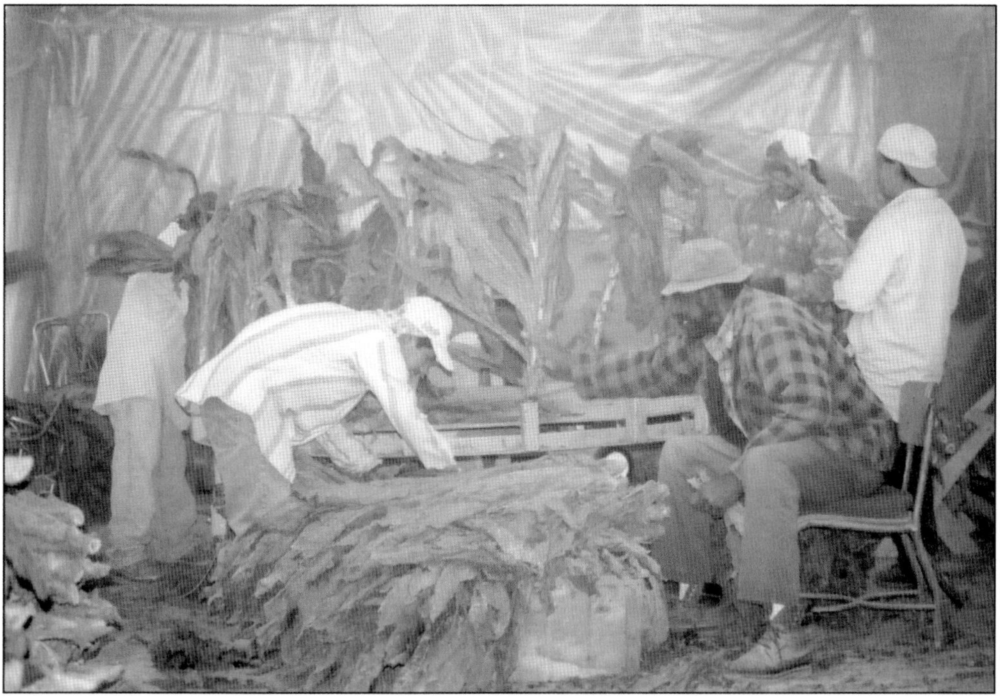

During the 1990s, the Griffiths farmed about 100 acres of tobacco. They purchased a tobacco stripping wheel to speed up stripping around 1990. The first man would strip off the seconds, then place the stalk on the wheel, where other men would strip their grade. Once the last grade was stripped, the stalk would be removed from the wheel. The graded leaves would be placed in a Handy Pack. (Griffith.)

Bernard Doepkens Sr. is showing off prize tobacco at the 1958 Anne Arundel County Fair held at Sandy Point State Park. He was named Young Farmer of the Year by the Annapolis Jaycees in January 1958 based on his work with soil conservation on his Harwood farm. He farmed with his father-in-law, Winterson Hardesty. (AASCD.)

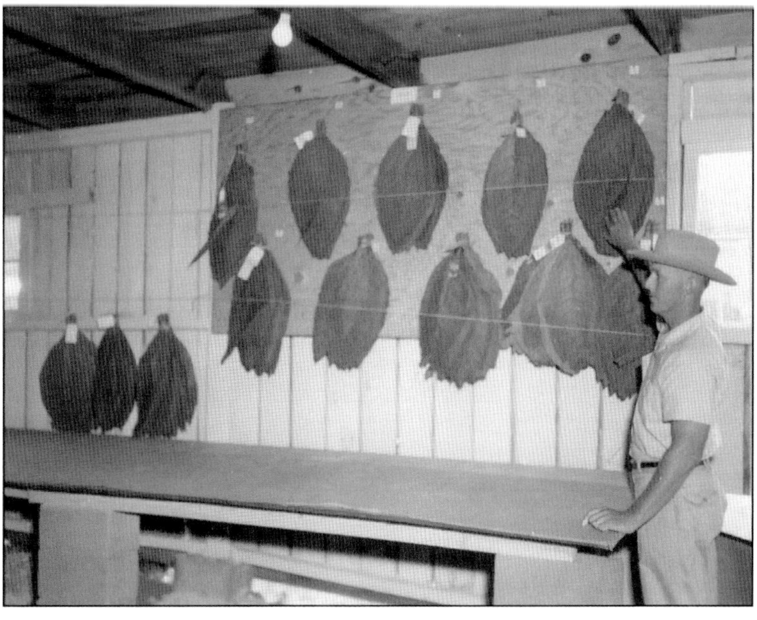

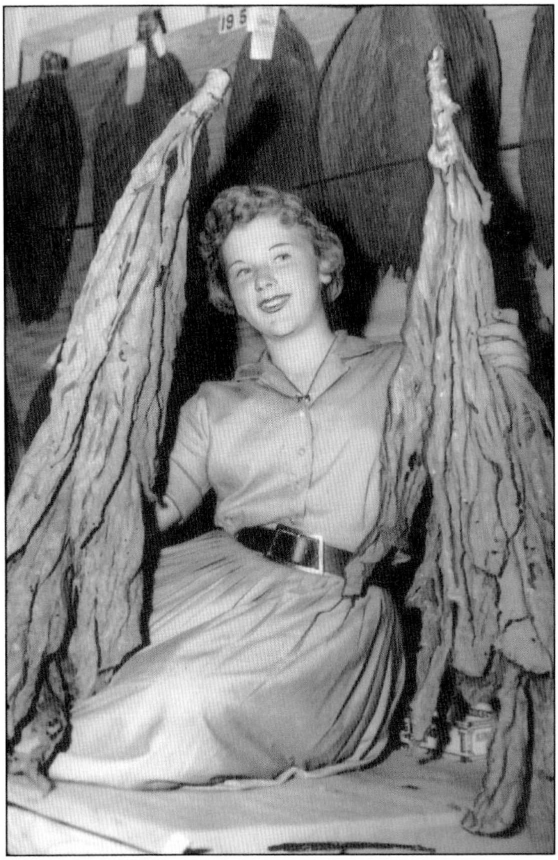

Margaret Nutwell poses with her prize stalks of tobacco in September 1959 at the Anne Arundel County Fair at Sandy Point State Park. Nutwell was a member of the West River 4-H Club. The 4-H is a national university-based youth developmental program for young people ages five to 18 and has been in existence since 1909. The local 4-H clubs are operated through the University of Maryland Cooperative Extension Service. (Hearst.)

Since colonial days, tobacco was packed in hogsheads for shipment to market. The wood used to make the hogshead usually was harvested from trees on the farm. Billy Watts, a laborer on Ivy Neck Farm, is working on cutting wood slats to make a hogshead around 1940. The standard hogshead size is 30 inches by 48 inches. (Colhoun.)

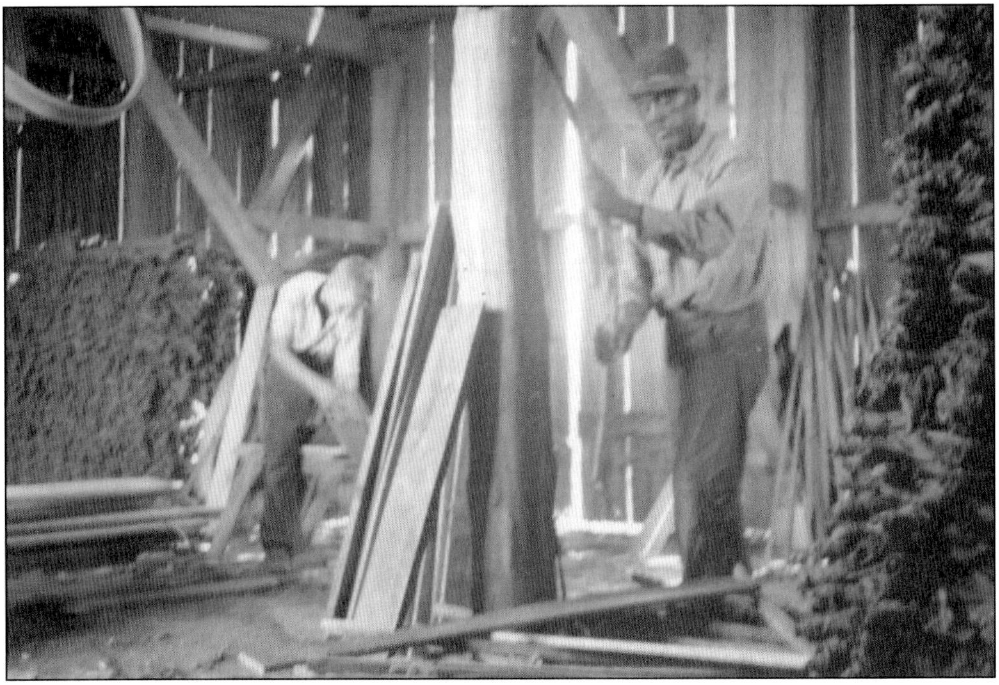

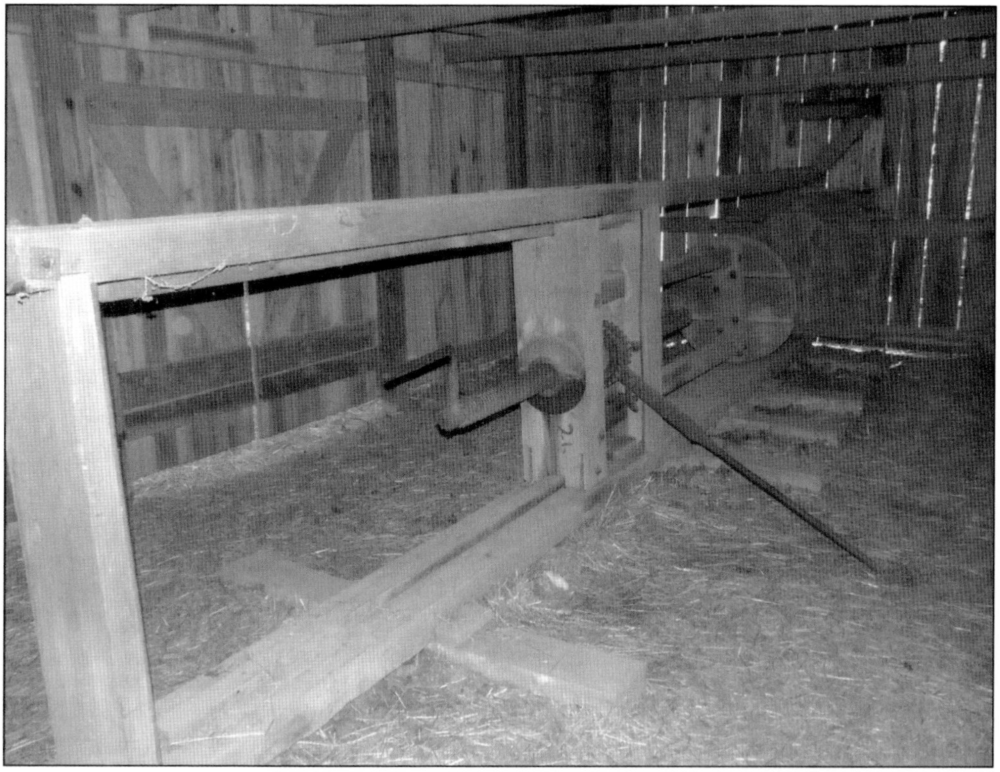

This is a tobacco prize that is used to pack the hands of tobacco for shipment. There are two basic styles: horizontal and vertical. In this horizontal prize, on display at the Smithsonian Environmental Research Center in West River in 2010, the hogsheads would be laid on their sides while being tightly packed. Once packed, they were ready for market and weighed nearly 1,000 pounds. (FHD.)

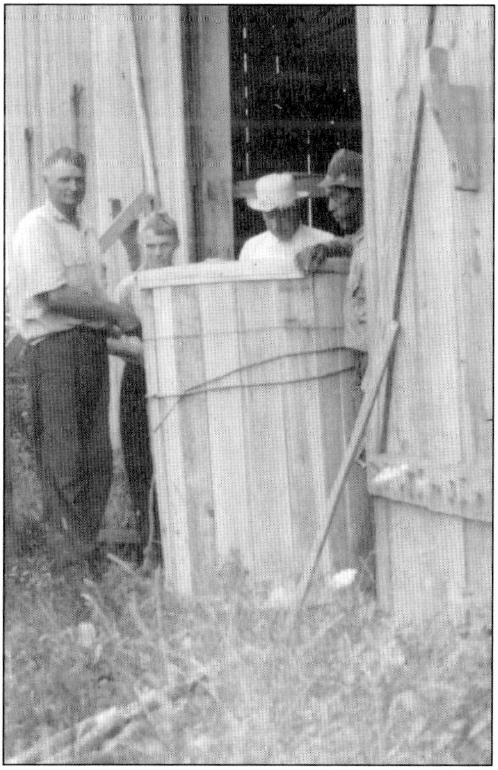

From left to right, Edward O'Neill, Mason O'Neill, Forbes Colhoun, and Billy Watts make a hogshead on Colhoun's Ivy Neck Farm in 1940. Not all farmers owned their own tobacco prize, which was a press used to pack tobacco into the hogsheads, so many had to take their tobacco to other farms in order to get their crop packed for market. (Colhoun.)

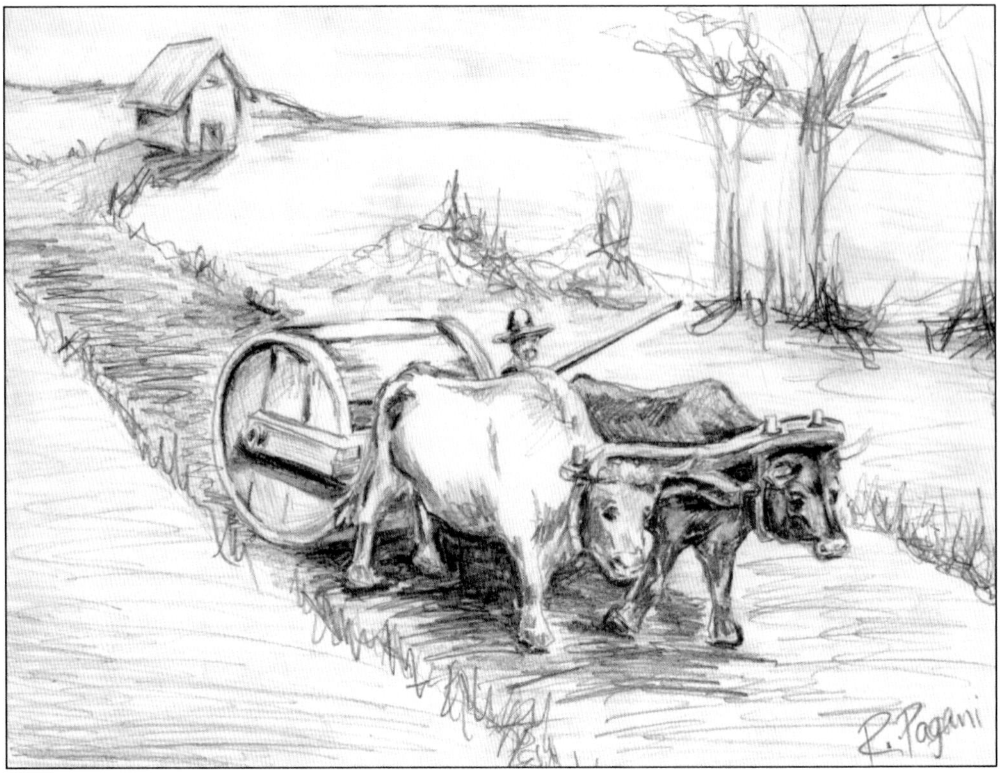

During colonial days, the packed hogsheads would be turned on their sides and rolled to a nearby wharf for shipment to England. The hogsheads were either rolled by hand or, as depicted in this 2010 sketch, attached to two oxen or horses and pulled down "rolling roads" to the closest wharf for shipment as described in Joseph L. Browne's *From Sotweed to Suburbia.* (Drawing by Rebecca Pagani.)

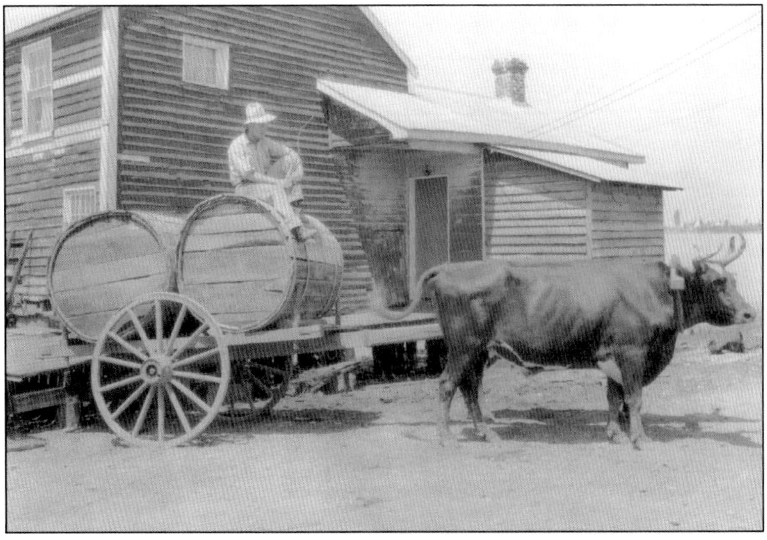

In this 1936 photograph, the tobacco hogsheads are being transported on a bullock cart and are waiting to be loaded onto a ship headed for Baltimore. (UMCP.)

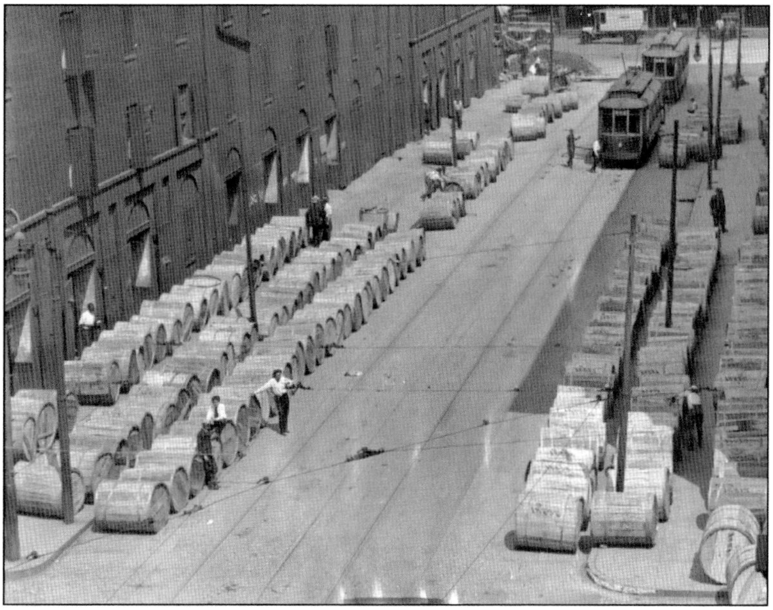

In August 1919, the crop of Maryland tobacco was so large—totaling 19,000 hogsheads—that it overflowed into the streets outside the tobacco warehouses on Charles and Conway Streets in Baltimore. Tobacco was last packed in hogsheads in the mid-1970s. (Hearst.)

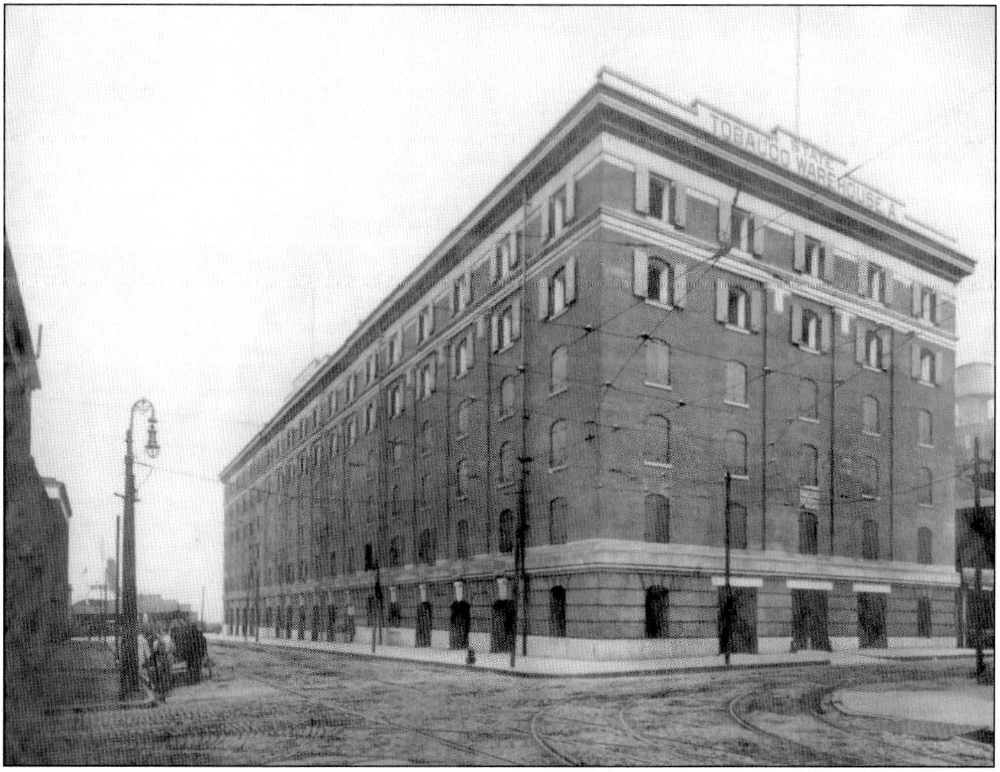

The Maryland State Tobacco Warehouse was located near the harbor in Baltimore and was the tobacco's last stop before its shipment overseas. There were several tobacco warehouses in Baltimore in 1920. Until 1939, all farmers sold their tobacco crop in hogsheads in one of these warehouses. (Hearst.)

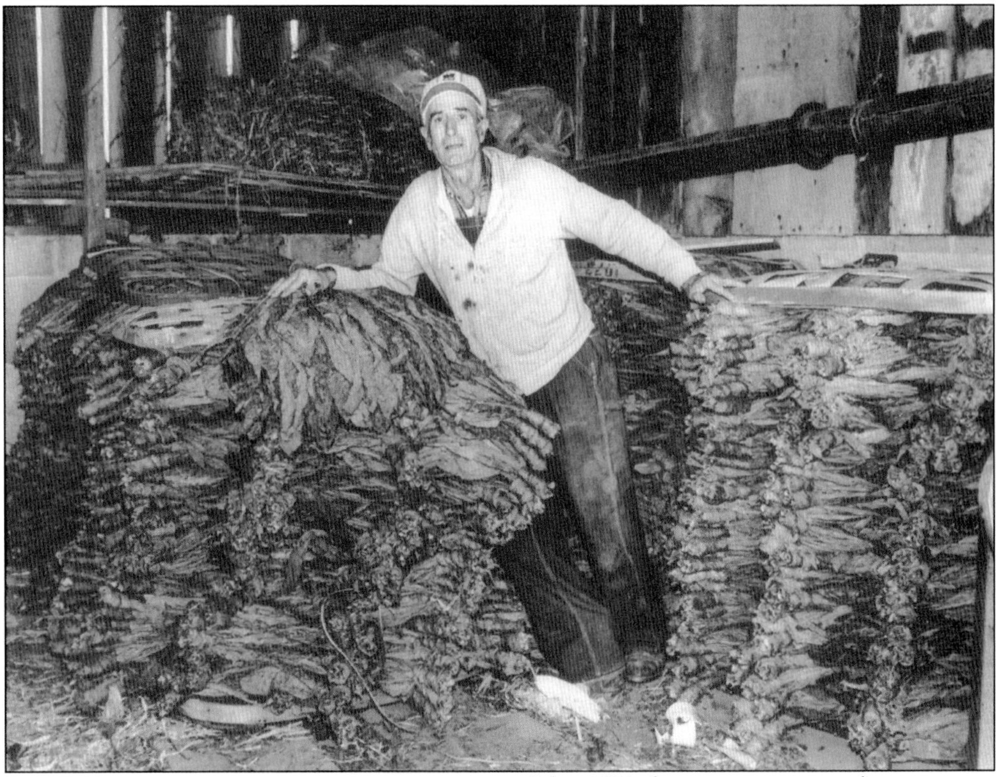

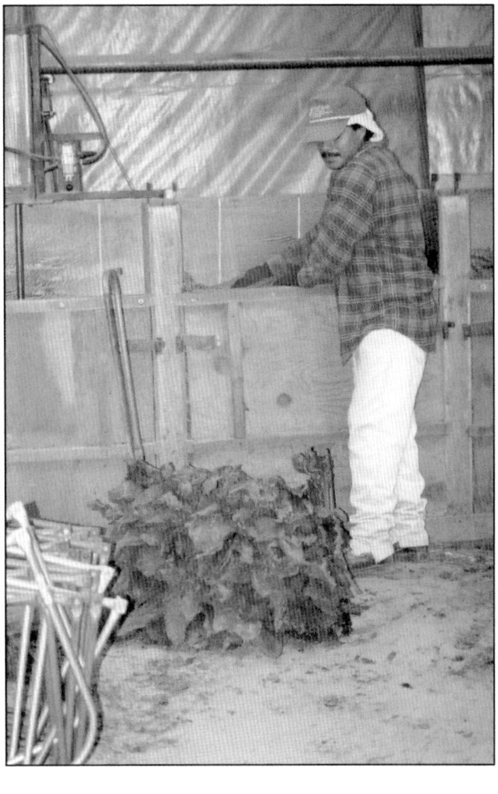

After 1939, farmers were given the option to pack their tobacco either in hogsheads or in a loose-leaf burden. This loose-leaf method became very popular, as all farmers could pack the tobacco on their own farms. Here, Oscar F. Grimes Jr. is standing next to some loose-leaf burdens stored in his barn in November 1982 ready for market in the spring. (Hearst.)

A worker on Earl Griffith's farm takes the graded leaves from the Handy Pack and packs the loose tobacco leaves into the tobacco baler in 2001. The tobacco leaves are added and pressed until the box is full. Strings are tied around these full bales of leaves and labeled with an identification tag, making them ready for market. This method of packing tobacco became popular in the 1990s. (Griffith.)

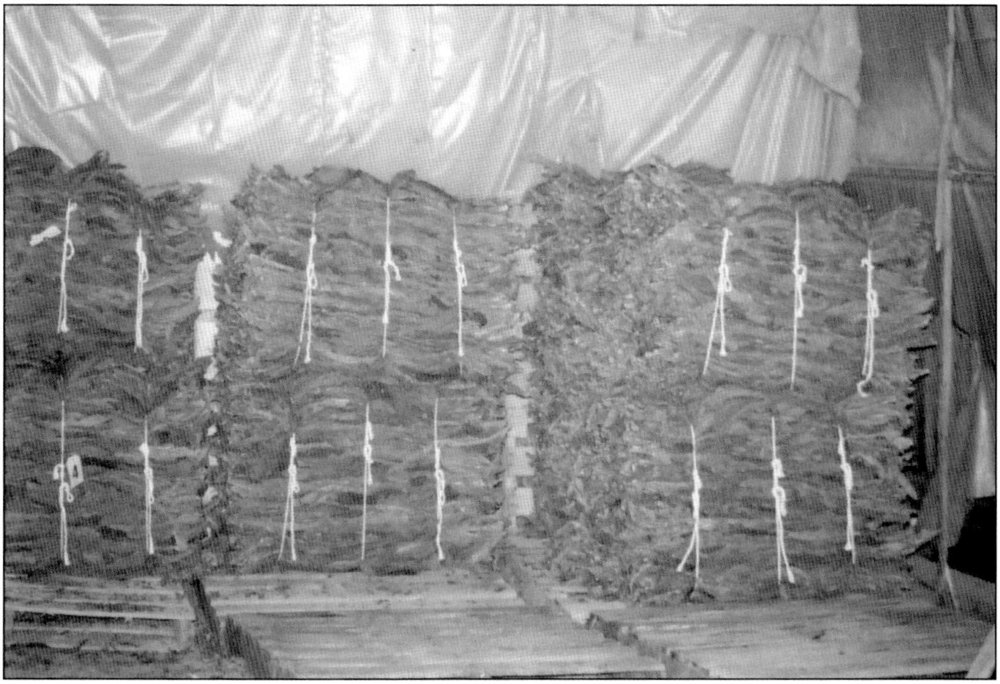

Once the tobacco stalks were stripped clean of the leaves, farmers usually spread them onto the tobacco fields for fertilizer. The Griffith farm baled up its tobacco stalks and sold them to tobacco companies to make some extra money in 2001. (Griffith.)

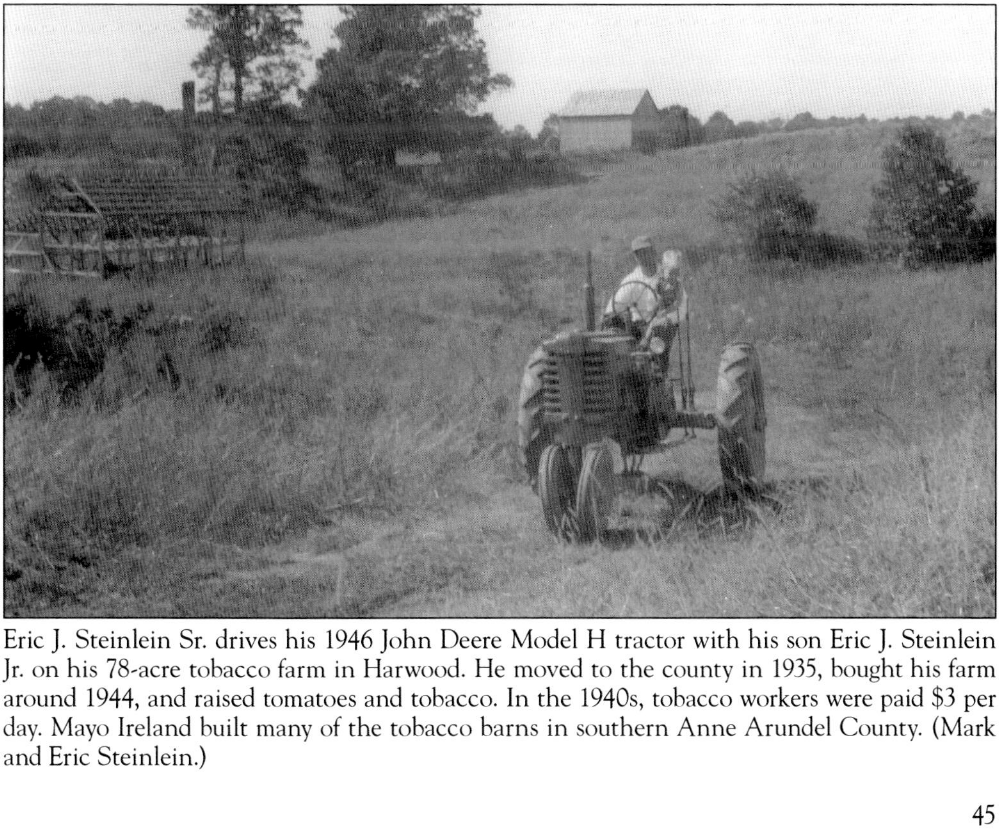

Eric J. Steinlein Sr. drives his 1946 John Deere Model H tractor with his son Eric J. Steinlein Jr. on his 78-acre tobacco farm in Harwood. He moved to the county in 1935, bought his farm around 1944, and raised tomatoes and tobacco. In the 1940s, tobacco workers were paid $3 per day. Mayo Ireland built many of the tobacco barns in southern Anne Arundel County. (Mark and Eric Steinlein.)

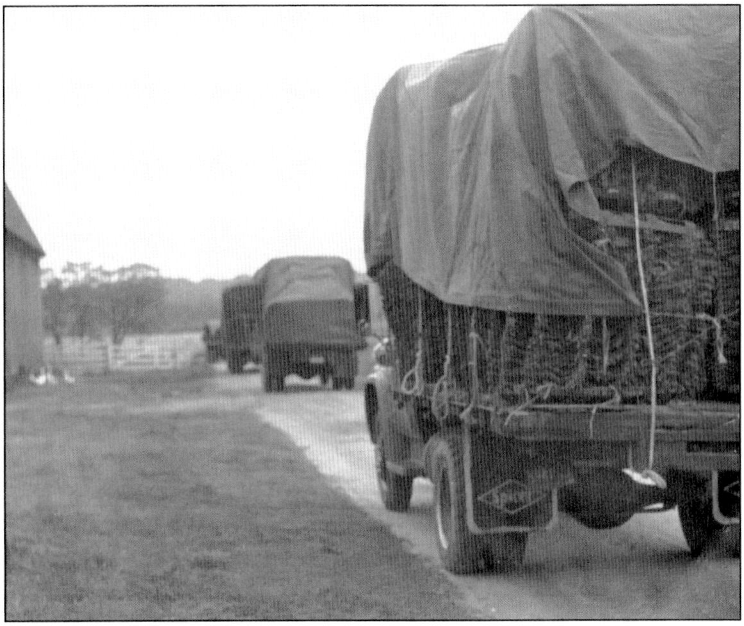

Once the loose-leaf tobacco auction houses opened in mid-April, it was time for the farmers to transport their burdens of tobacco to market. Here trucks are loaded up on the Charles D. Pettebone farm near Annapolis in 1970. If farmers didn't have trucks to transport their crop, the tobacco auction houses would pick up the crop and transport it to market. (Roxanne Pettebone McKagan.)

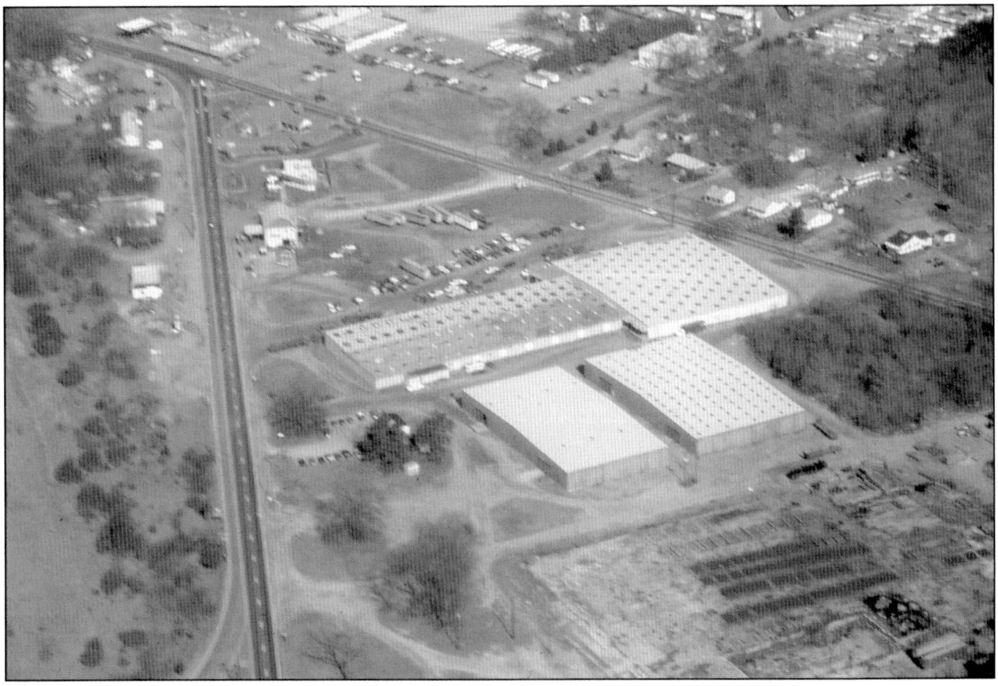

An April 1987 aerial view shows the Triangle Tobacco Warehouse in Wayson's Corner. This was the only loose-leaf tobacco warehouse in the county and was operated by the Maryland Tobacco Growers' Association. Other loose-leaf tobacco warehouses that were used by farmers were in nearby Upper Marlboro, Maryland. The warehouse was the site of a yearly tobacco auction until it closed in 1998. (AASCD.)

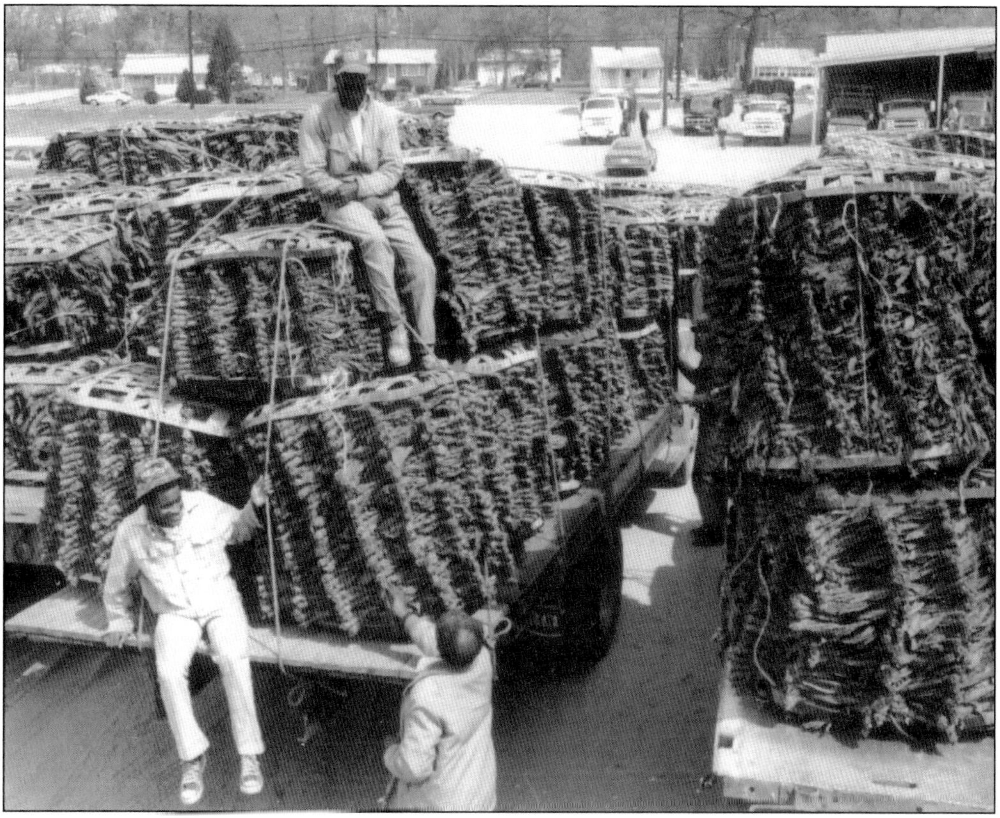

Loaded trucks are parked outside the Triangle Warehouse in 1978, waiting to be unloaded by the workers. The loose-leaf markets were open from mid-April until June. By 1951, more than 92 percent of all Maryland tobacco markets were using the loose-leaf system. (Sun.)

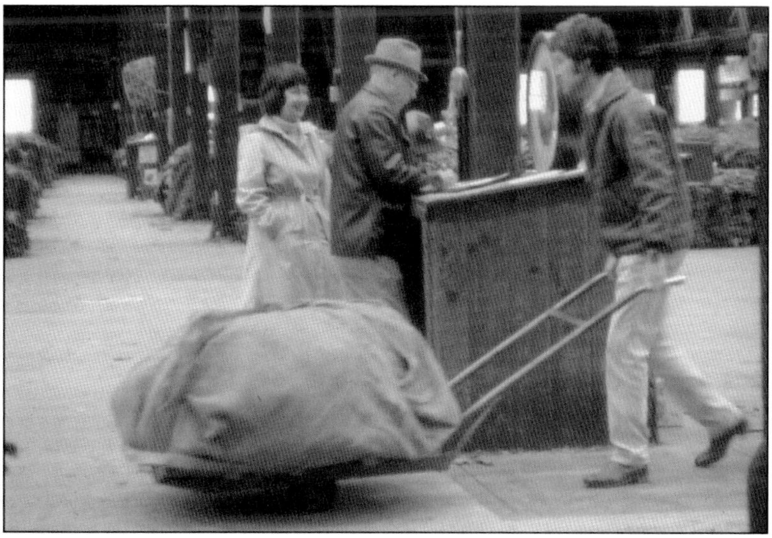

Once the tobacco burdens are unloaded, they are weighed on a scale, tagged, and transported to the auction floor, as seen here in 1981 at the Edelen Brothers Tobacco Warehouse in Upper Marlboro. (Salyers.)

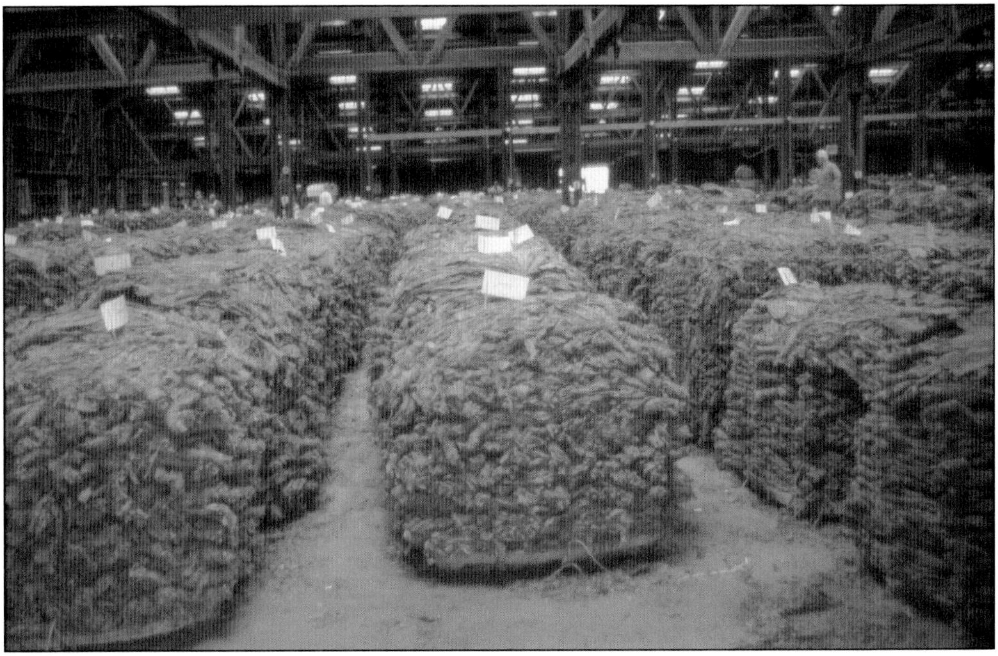

The burdens from the different farmers are lined up and awaiting the auctioneer. Each burden is labeled with the farm information, the grade of tobacco, and the weight of the burden. This photograph was taken in the Edelen Brothers Tobacco warehouse in Upper Marlboro in Prince George's County in 1981. (Salyers.)

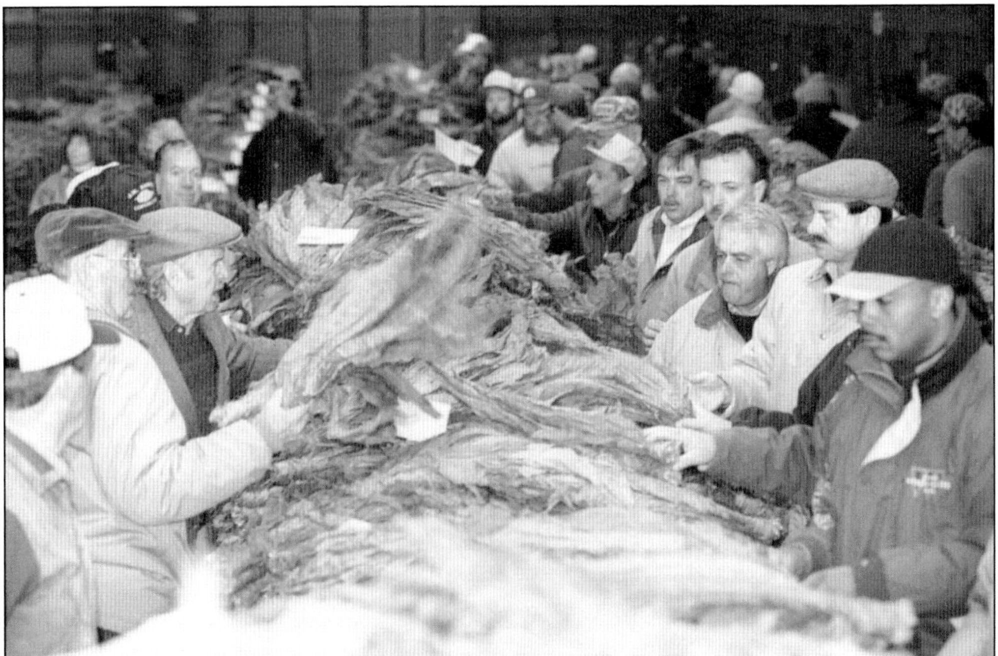

Triangle Warehouse manager Bernard Doepkens Sr. (left), accompanied by an auctioneer and the tobacco company buyers, canvasses the rows of burdens and bids on tobacco around 1980. The tobacco is sold to the highest bidder. (Capital.)

The tobacco auctioneer, Bob Cage (left), chants out the prices as prospective buyers representing the major tobacco companies inspect the tobacco in 1991. In a matter of minutes, the 19 months of preparation, growth, and harvesting of the farmer's tobacco crop is finished, and he is paid for his crop the same day. (Sun.)

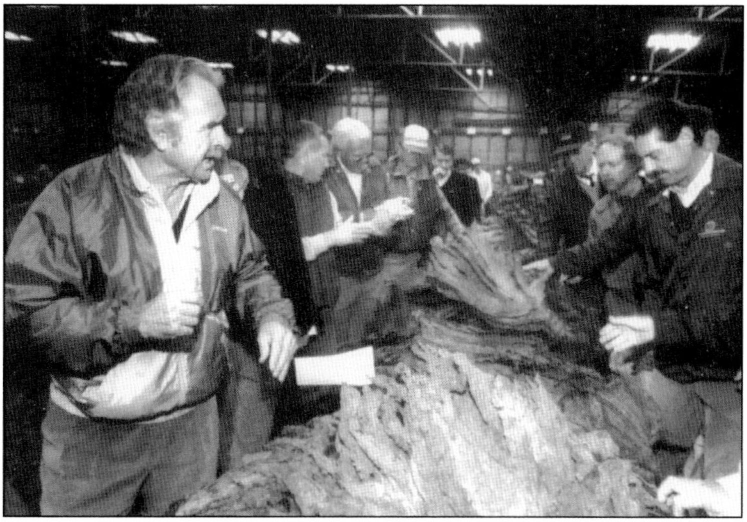

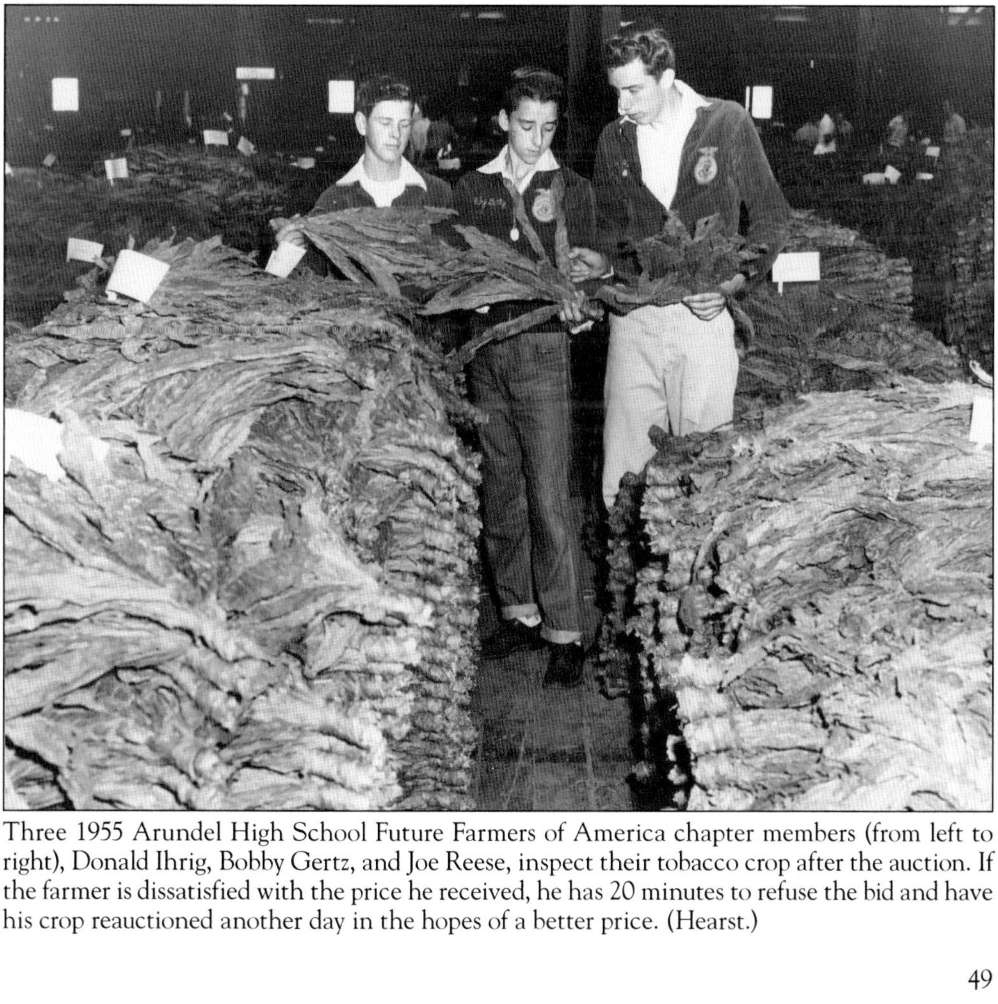

Three 1955 Arundel High School Future Farmers of America chapter members (from left to right), Donald Ihrig, Bobby Gertz, and Joe Reese, inspect their tobacco crop after the auction. If the farmer is dissatisfied with the price he received, he has 20 minutes to refuse the bid and have his crop reauctioned another day in the hopes of a better price. (Hearst.)

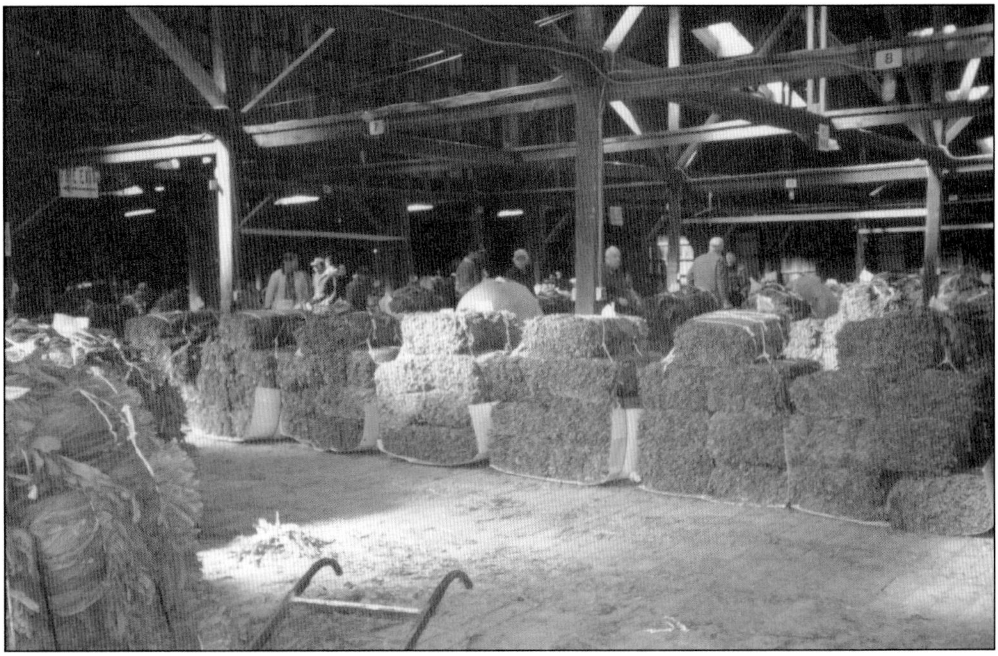

Bales of tobacco from the William Gertz farm await sale at the Planters' Auction House in Upper Marlboro around 1990. These bales are smaller and easier to transport than the larger 200-pound burdens. (Gertz.)

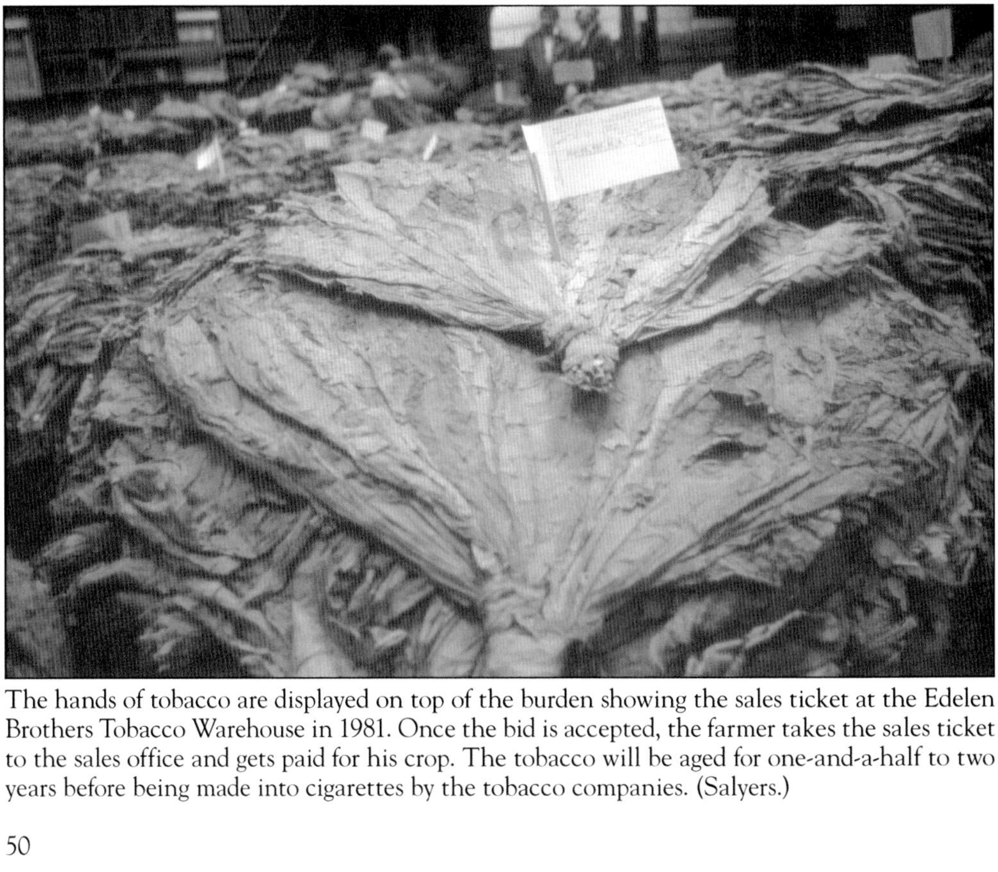

The hands of tobacco are displayed on top of the burden showing the sales ticket at the Edelen Brothers Tobacco Warehouse in 1981. Once the bid is accepted, the farmer takes the sales ticket to the sales office and gets paid for his crop. The tobacco will be aged for one-and-a-half to two years before being made into cigarettes by the tobacco companies. (Salyers.)

Two

Truck Farming

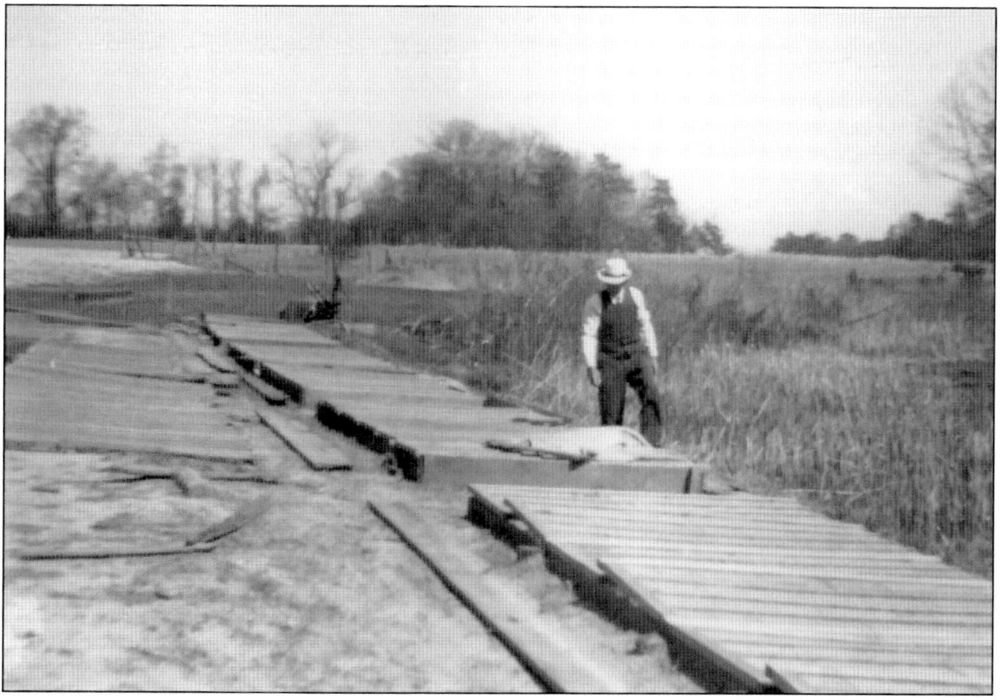

Plants for the north county truck farms were started under glass in February and March to be ready for the field when all frost was gone by May 10. The Charles Hancock Calvert farm around 1930 had about 75 sashes available and used 50 to 60 each year for starting cantaloupe and tomato plants in cans to be planted into the fields after frost. (Calvert.)

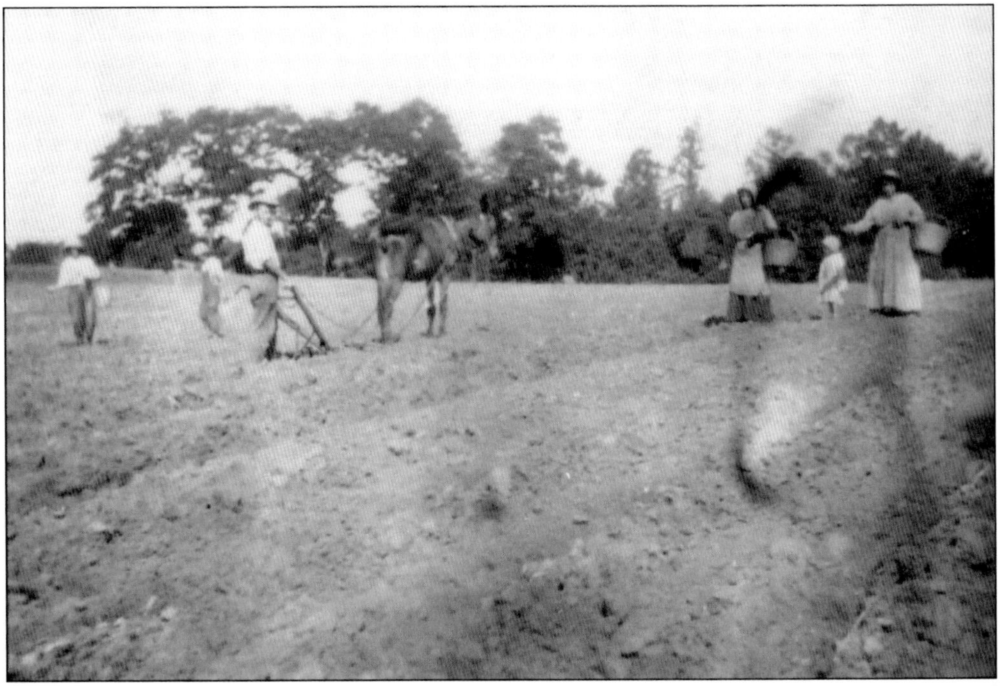

John Gertz and family plant potatoes on their Crownsville farm in 1916. The crop would be used both to feed the family and to be sold in markets in Baltimore. (Gertz.)

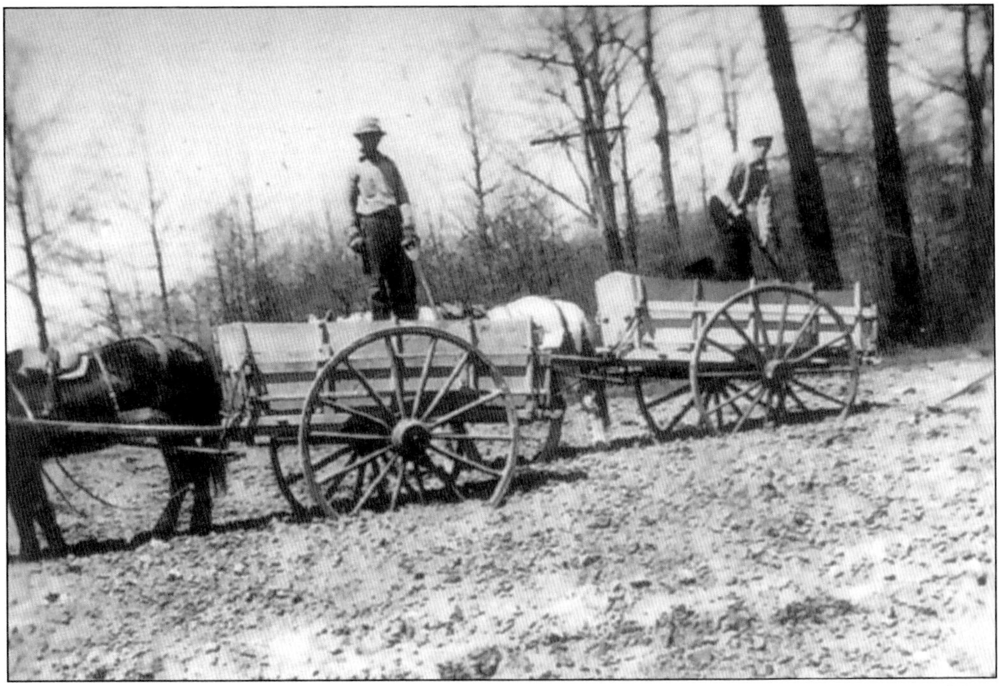

In 1931, two one-horse carts place manure in the hills to fertilize cantaloupes. Rows, spaced five feet apart, were opened with a single-horse boss plow. The compost manure was placed at four-foot intervals in the hills and covered or "listed up" with the same plow. Then children would drop the plants for the adults to plant in the ground. (Calvert.)

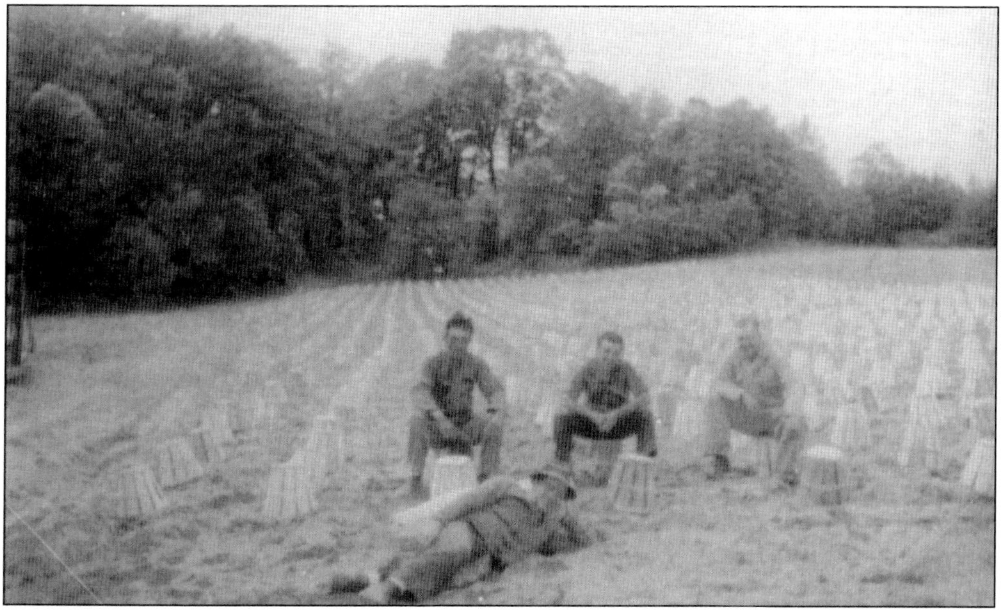

Workers on the Max Blob farm in Jessup around 1942 cover up their newly planted tomato crop with half-bushel baskets to protect the tender plants from the danger of frost. Pictured are, from left to right, unidentified, Max Blob, John Eggerl Jr., and John Eggerl. Max emigrated with his family from Vogenberg, Germany in 1896 and settled on 150 acres of open farmland in Jessup. (John Eggerl Jr.)

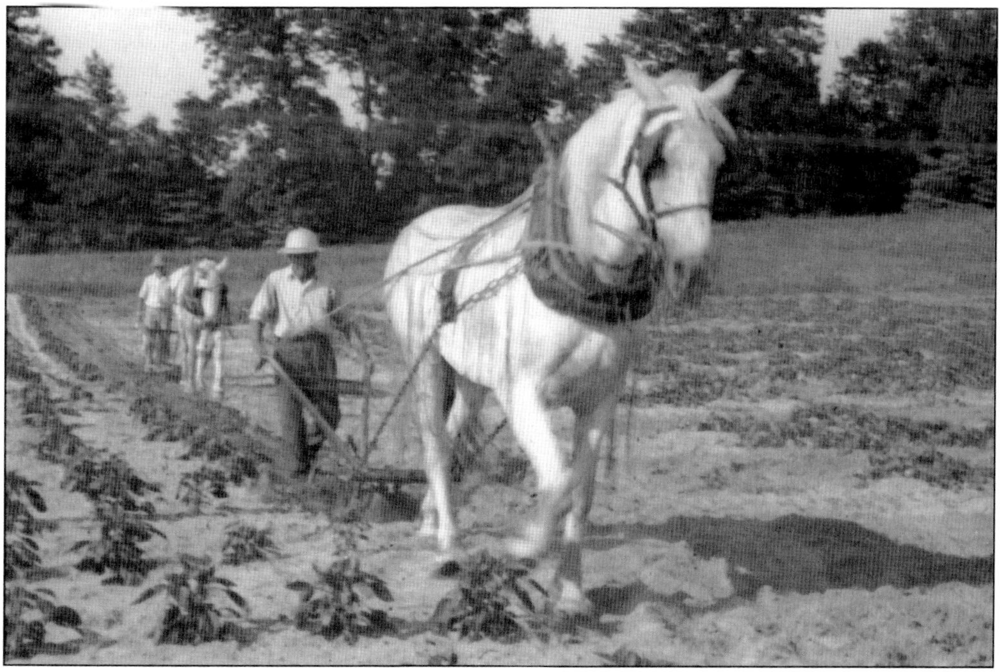

Charles H. Calvert, with farm horse Bill in early July 1940, cultivates pepper plants. The second cultivator and horse are Edwin Calvert (age 14) with old Nell, learning to cultivate growing plants. The crops in the field to the right are cantaloupes, one of the best money crops on their produce farm. (Calvert.)

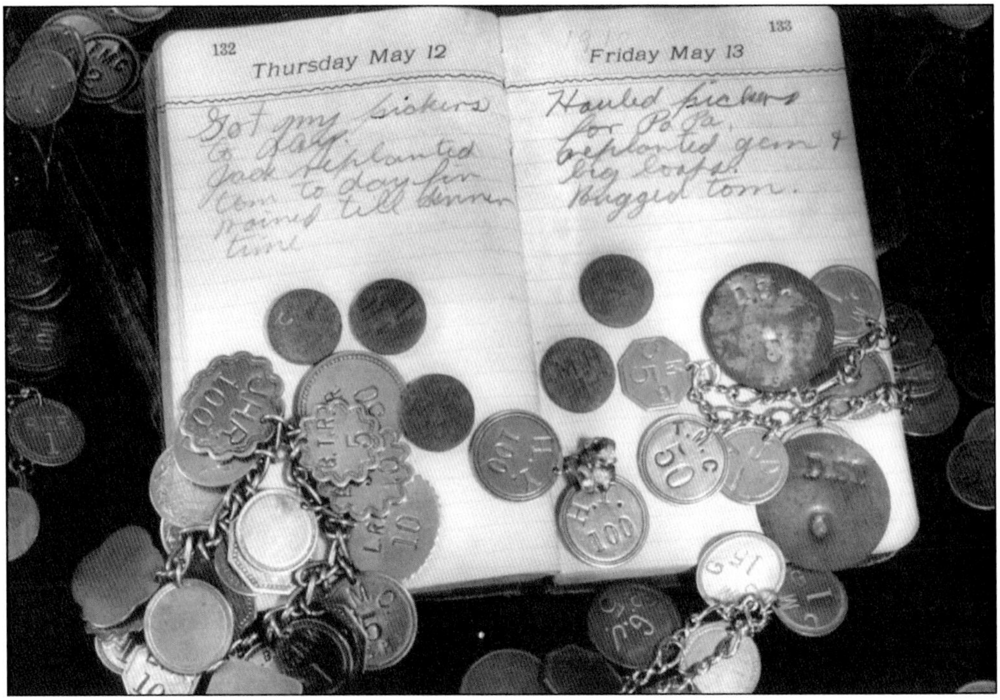

Pickers' checks were an important part of the truck farming industry in Anne Arundel County. Migrant laborers who worked in the field harvesting fruit and vegetable crops were paid these pickers' checks when they turned in their baskets of produce. At the end of the day, the checks were redeemed for money. The checks were stamped with the initials of the farm. (AACHS.)

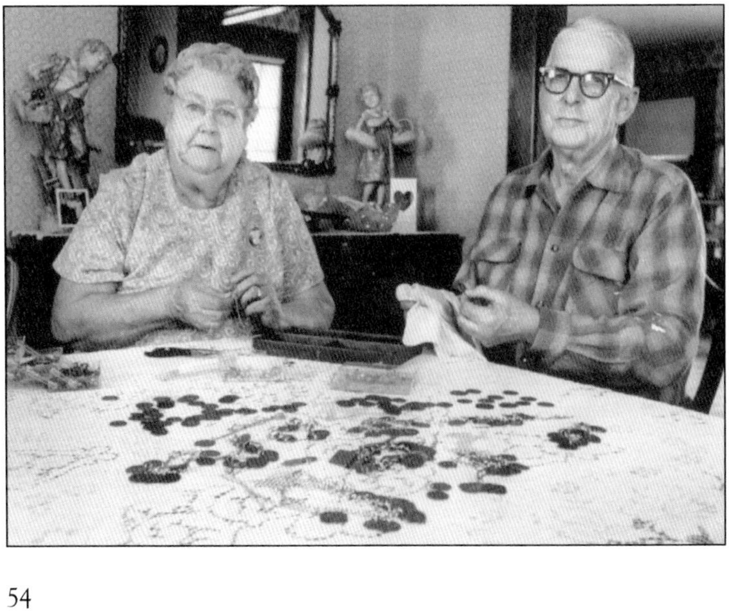

Ora Pumphrey Smith and her husband, John Smith, make bracelets from pickers' checks from local farms in 2000. John had a 75-acre truck farm in Severn on which he raised cantaloupes, tomatoes, peas, and beans. Eventually tobacco and sweet potatoes became the staple crops rather than vegetables. (AACHS.)

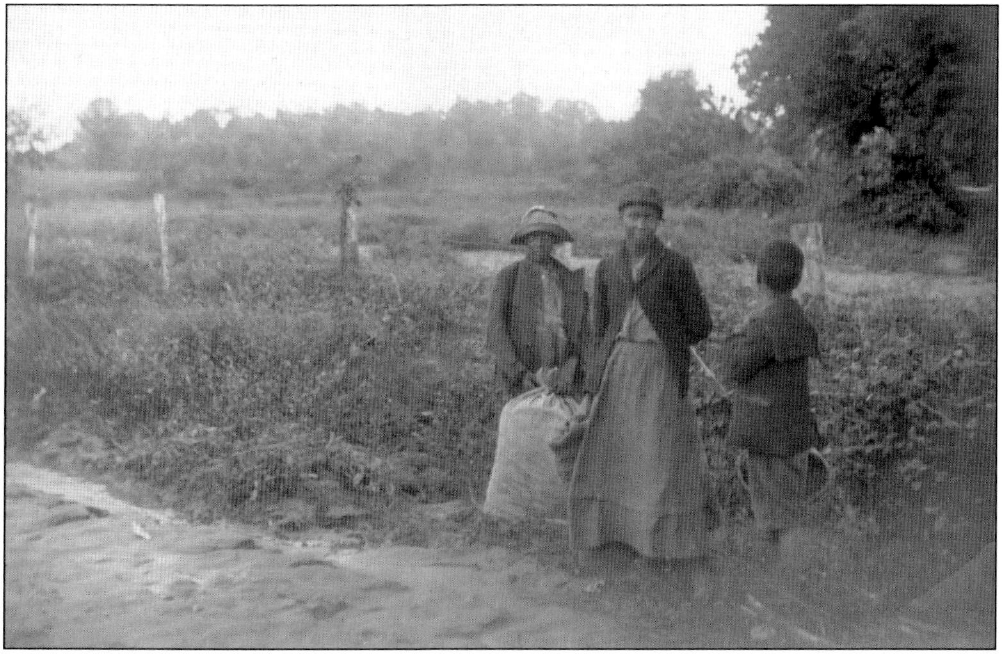

Pea pickers work in the field at Cherry Branch on the West River near Ivy Neck Farm. At the beginning of the 20th century, many south county farms raised vegetables in addition to tobacco. The vegetables were loaded onto steamers like the *Emma Giles* at wharfs in Taylorsville (now Riva) and throughout southern Anne Arundel County for transport to markets in Baltimore. (Colhoun.)

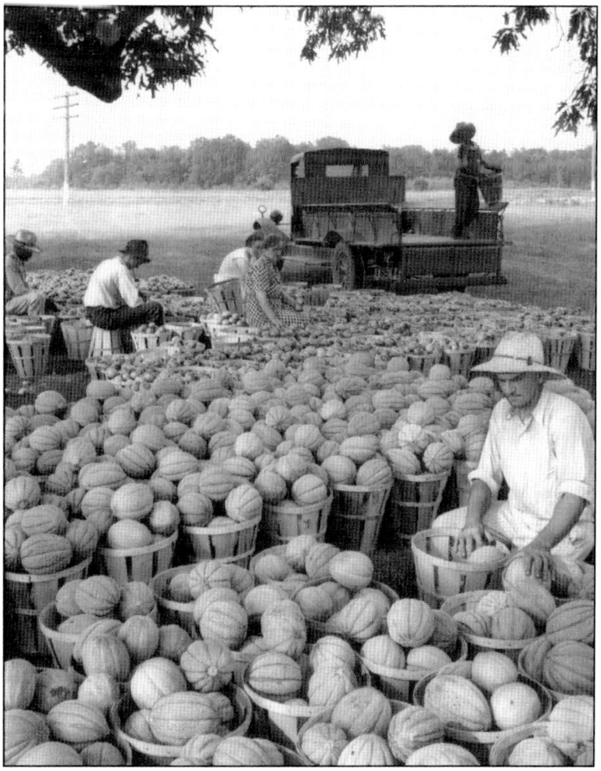

According to Willard Mumford's book on truck cropping, the Long John variety of cantaloupes was originated by John Shipley on his farm near Bayside Beach around 1910. The cantaloupes were one of the green-meat melons that were developed in Anne Arundel County and are still grown by farmers in the county. This 1936 photograph was taken on an unidentified Anne Arundel County farm. (Sun.)

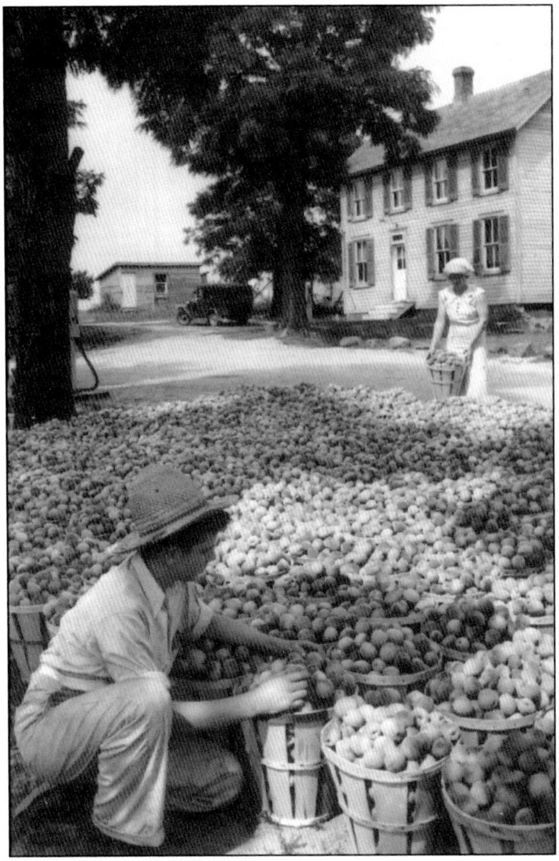

Here is a farmer in the northern part of the county with half-bushels of peaches ready for market. The George and Henry Schmidt family were large growers of peaches in the Pasadena area. This 1936 photograph was taken on an unidentified Anne Arundel County farm. (Sun.)

Amelia and Ida Kinder sit on top of a load of squash and sweet potatoes being transported to market from the Gustave Kinder farm around 1930. The Kinder family emigrated from Germany and bought their farm along Jumpers Hole Road in Millersville in 1898. They slowly acquired more land, and from the early 1900s to the late 1930s, they grew vegetables and fruits to sell in the Baltimore markets. (Kinder.)

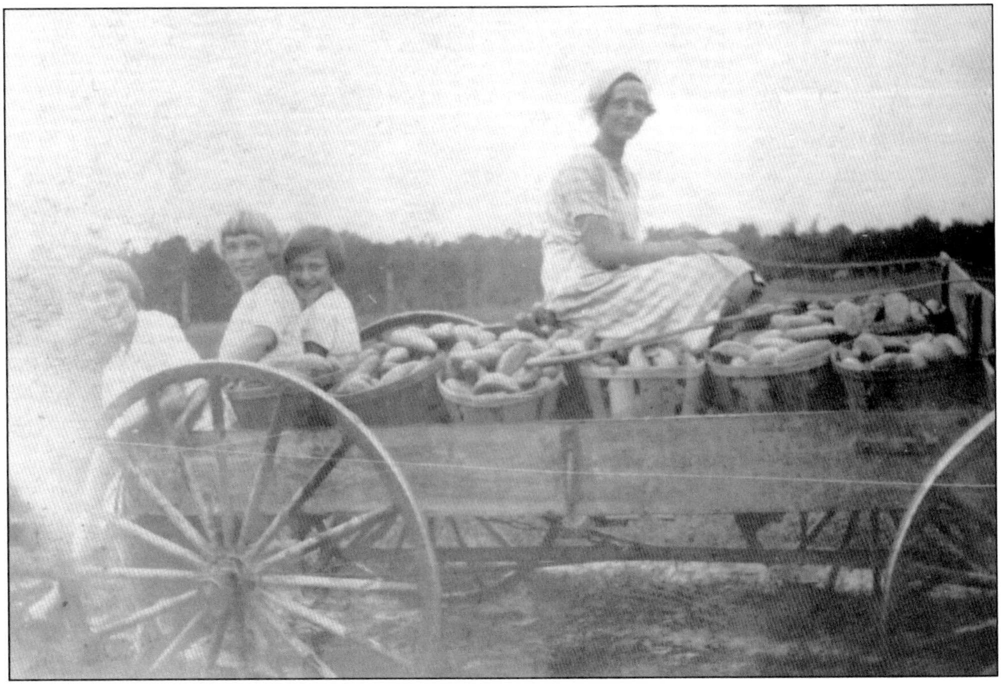

The William J. Schmidt and Sons Stone House Stand was located in Pasadena on Schmidts Lane off Mountain Road in 1933. They sold fresh fruits and vegetables grown on their farm from 1920 until 1962. (Kinder.)

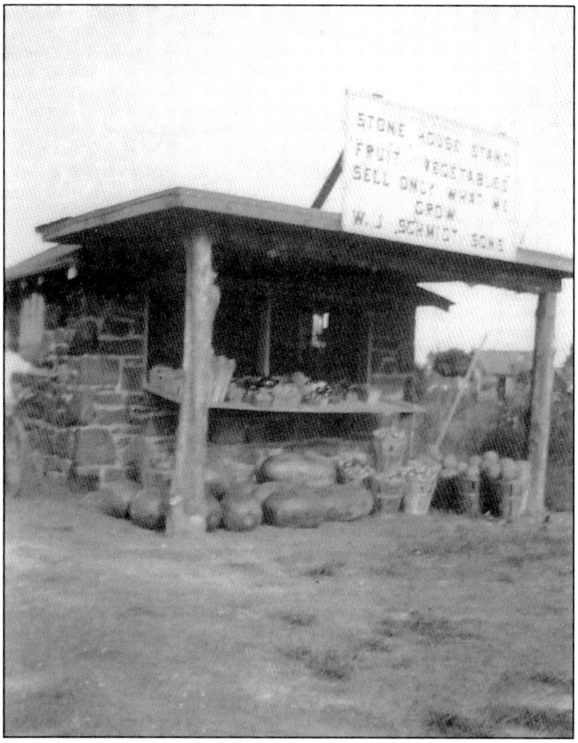

When the Crownsville State Mental Hospital was formed in 1910, the Boswell-Garrett-Hatch farm in Crownsville was purchased and farmed by the hospital's patients, as seen in this 1915 photograph. The farm was active until it closed in the late 1960s. (MSA.)

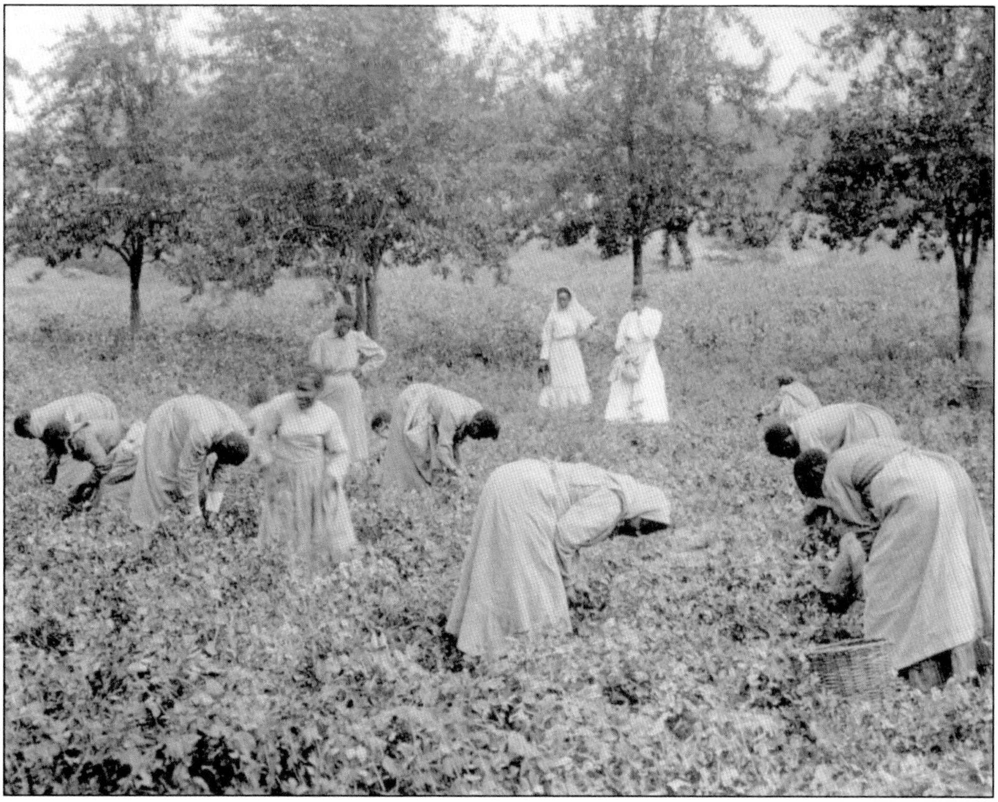

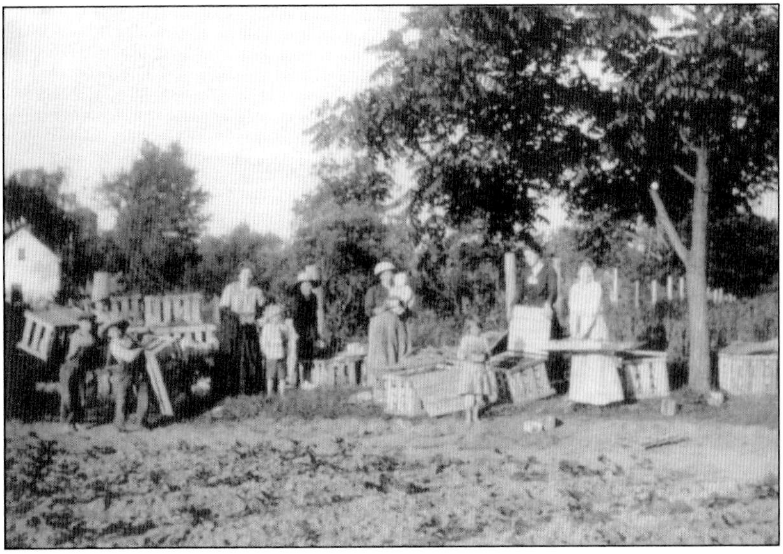

Polish pickers from Baltimore pick produce on the farm of Henry Emil Gertz in Crownsville in 1914. The produce was packed in 60-quart crates for shipment to markets in Baltimore. (Gertz.)

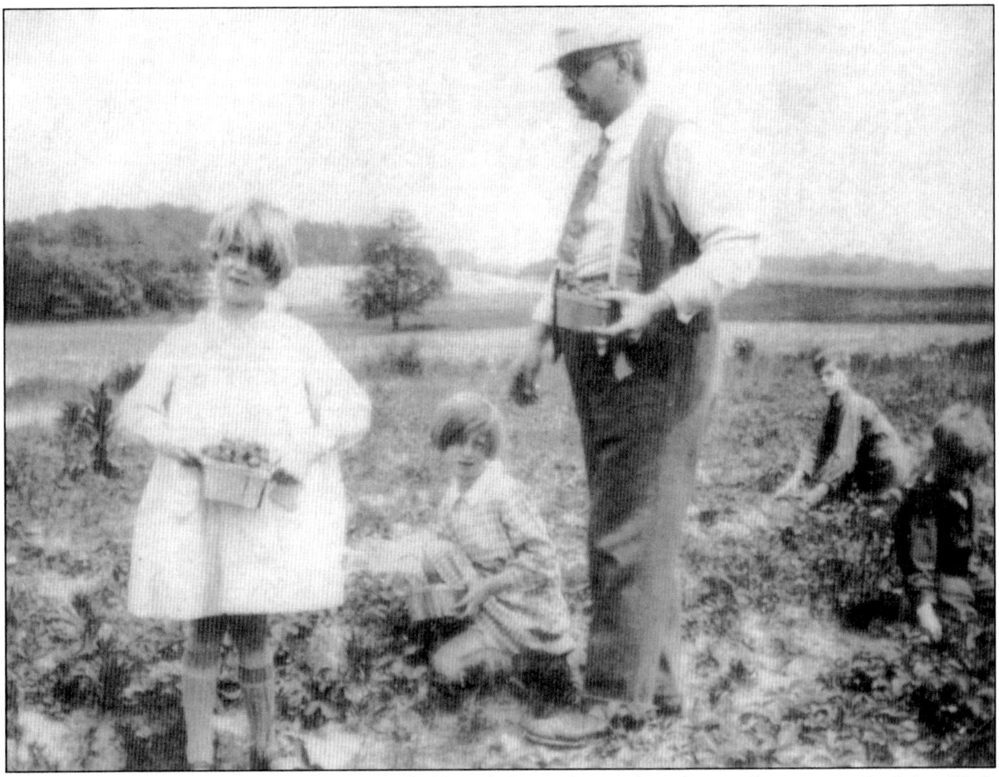

Mr. Hovestadt and his daughters, Marie (left) and Anna (center), pick strawberries on the Max Blob farm in Jessup around 1940. To the right are (from left to right) Max and Gene Blob. According to Mumford, by 1910, Maryland grew more strawberries than any other state, and Anne Arundel County led the counties with 3,937 acres planted. Strawberry festivals became popular fundraisers for local church groups. (John Eggerl Jr.)

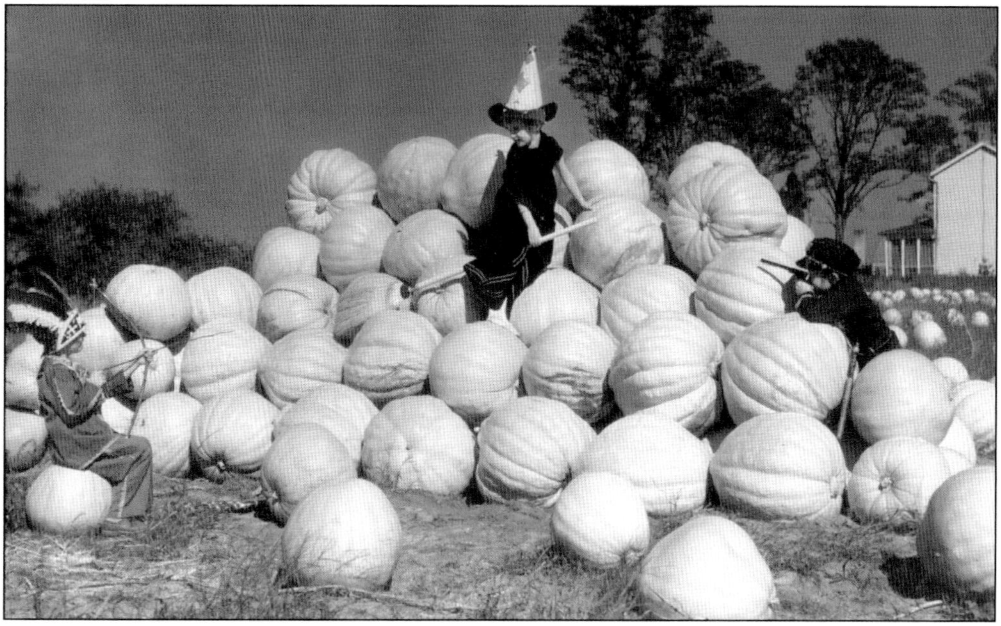

In October 1933, A. Aubrey Bodine took this picture of (from left to right) William M. Schramm II, Emma Schramm, and Louis F. Schramm (ages six, five, and eight) playing on a pile of pumpkins. Their grandfather, Louis Schramm Sr., grew pumpkins on his farm in Pasadena and won first prize at the Maryland State Fair. (Sun.)

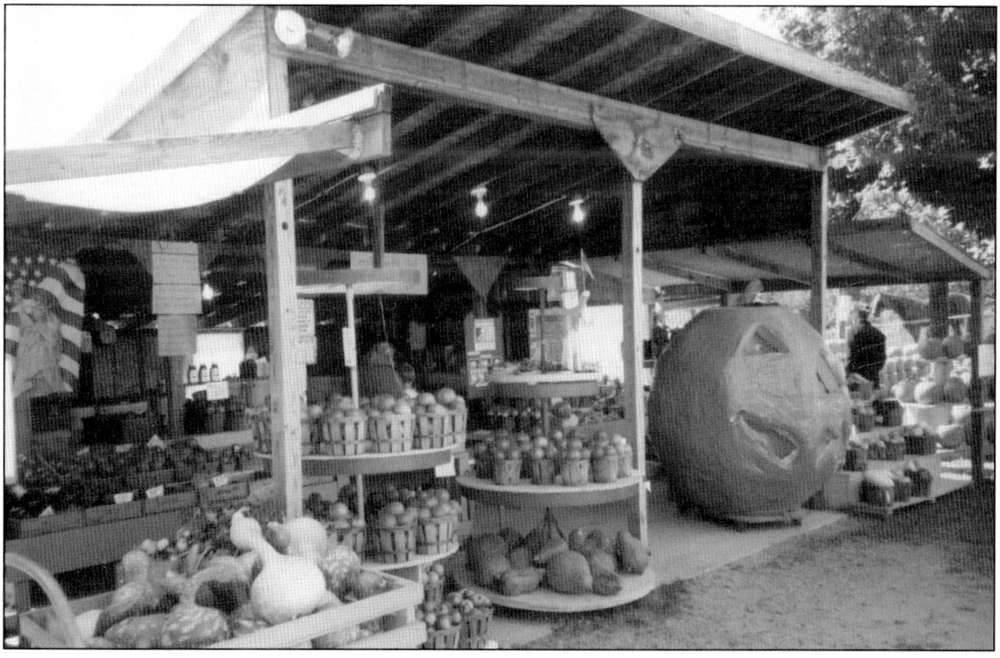

In 1997, the Schramm family sold most of their Pasadena farm, including the original farm stand. In this c. 1990 photograph, a smaller stand on their farm was used to sell their produce. When a customer purchased a pumpkin for $1 or more, his or her child could reach a hand into "Mr. Pumpkin" (the large pumpkin) and win a grab bag containing pieces of penny candy and a small toy. (Schramm.)

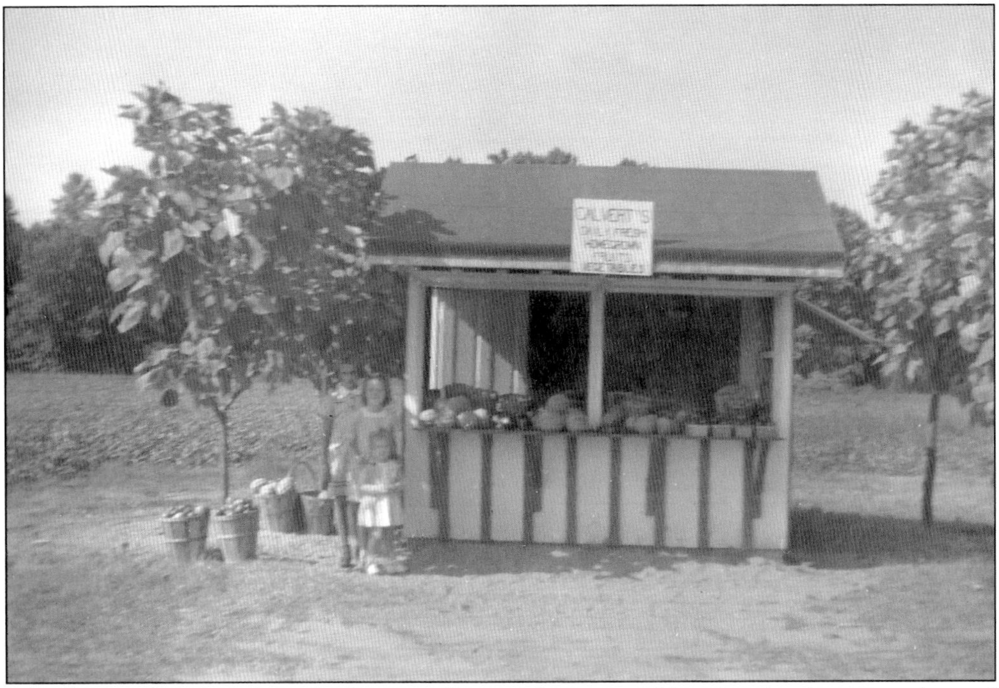

After bathing beaches along the Patapsco River and year-round housing developments changed the area, August Calvert built a farm stand to sell produce on his farm instead of taking it to Baltimore markets. Pictured from left to right are Annabel Calvert, daughter of Charles H. Calvert, and her two nieces, Eleanor and Diana Calvert, who lived across the road from the stand in Pasadena around 1940. (Calvert.)

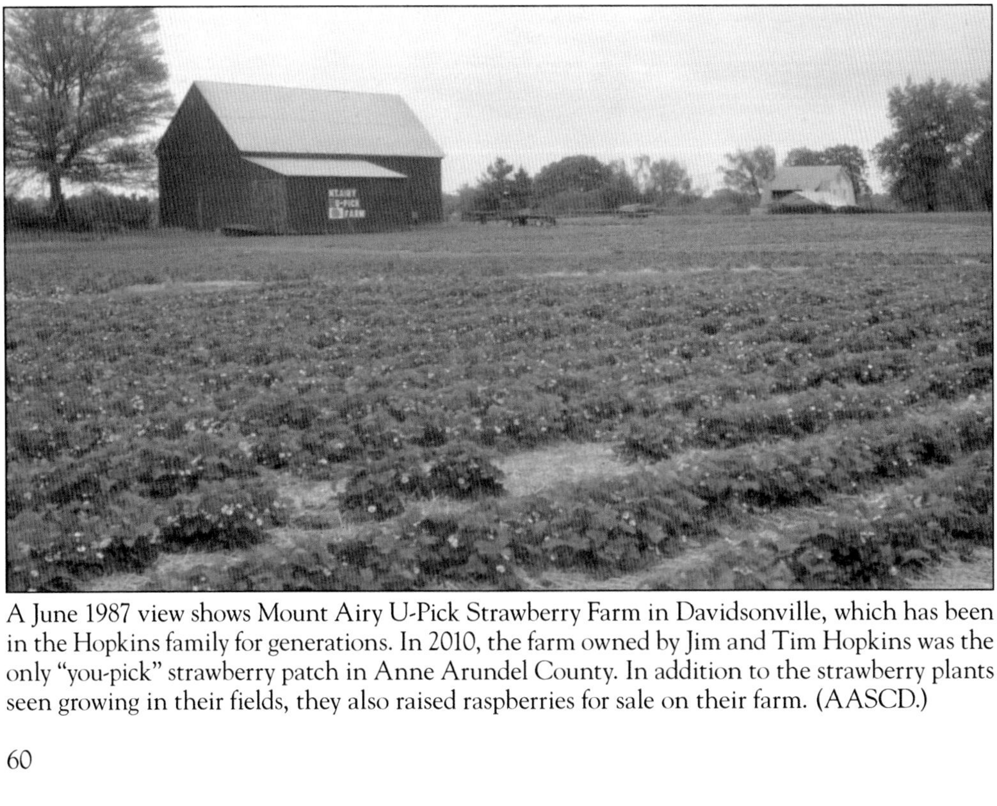

A June 1987 view shows Mount Airy U-Pick Strawberry Farm in Davidsonville, which has been in the Hopkins family for generations. In 2010, the farm owned by Jim and Tim Hopkins was the only "you-pick" strawberry patch in Anne Arundel County. In addition to the strawberry plants seen growing in their fields, they also raised raspberries for sale on their farm. (AASCD.)

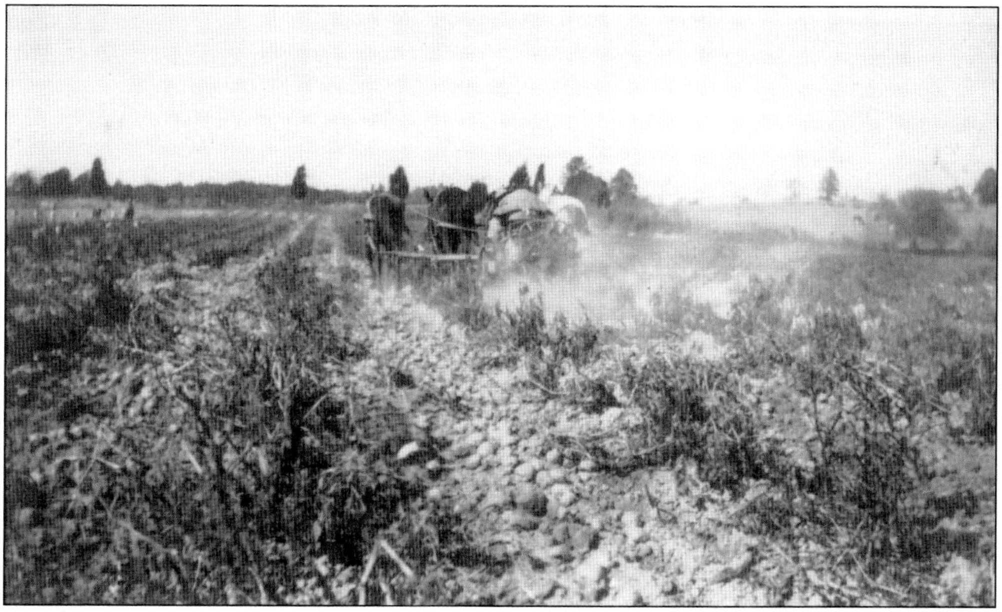

C. Addison Hodges harvests potatoes around 1913 with a team of horses on his Mount Elm farm in Davidsonville. Once the potatoes have been dug, they will be packed into baskets and taken to markets in Baltimore to be sold. In 1923, the farm was sold to Louis Hall and converted into a dairy farm. (Mary Jay Davis.)

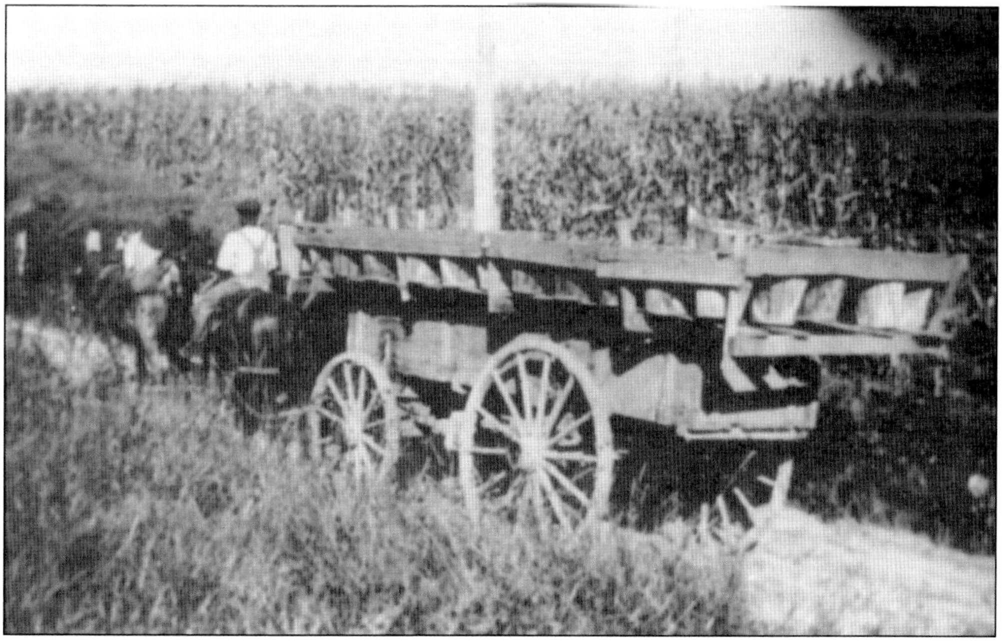

According to William Doepkens in his book on Middle Plantation, this photograph shows Joe Smith in 1913 riding a four-mule team hauling a 100-bushel load of tomatoes from Addison Hodges's Mount Elm farm to Taylorsville, where it would be loaded onto a boat for transport to markets in Baltimore. Taylorsville is now known as Riva. (Mary Jay Davis.)

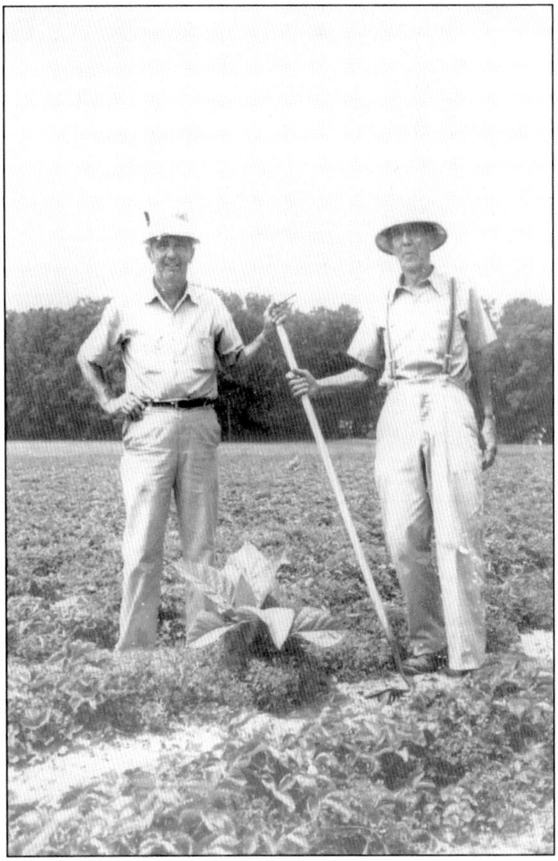

Charles D. Pettebone (left) and J. Elliott Pettebone (right) are pictured in one of the strawberry patches on their 120-acre farm in 1978. In the late 1970s, they started raising pick-your-own strawberries on their farm. The farm originally raised cattle, tobacco, and truck crops like tomatoes, cantaloupes, and watermelons. The farm has been in their family since the 1600s. (Roxanne Pettebone McKagan.)

Customers pick fresh strawberries on the Charles D. Pettebone farm on the Broadneck Peninsula near Annapolis in 1984. About eight acres of strawberries were grown, with four acres being picked on a two-acre rotation. A portable farm stand was moved to the patch that was currently being picked. (Roxanne Pettebone McKagan.)

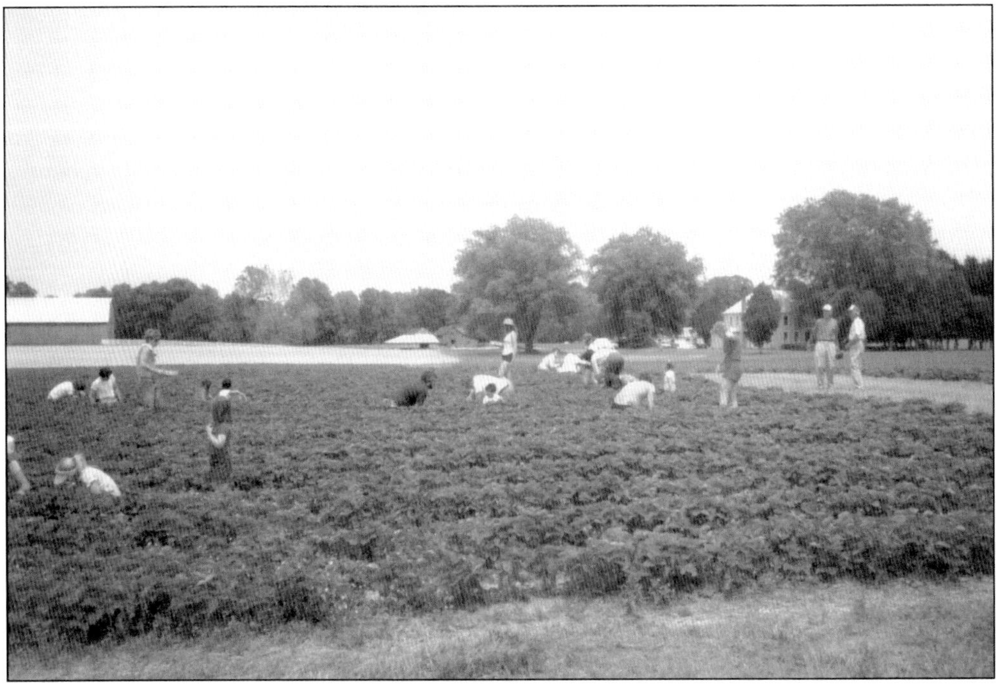

Martin and Hilda Zehner sell their produce at the farmers' market on Riva Road near Annapolis around 1990. Martin Zehner operates a 114-acre farm in Davidsonville and raises an assortment of grain and vegetable crops as well as cattle. His dad, Martin Zehner Sr., immigrated to the United States from Bavaria, Germany, in 1927 and purchased a farm in Davidsonville in 1936. (Photograph by Marion E. Warren, © M.E. Warren Photography, LLC.)

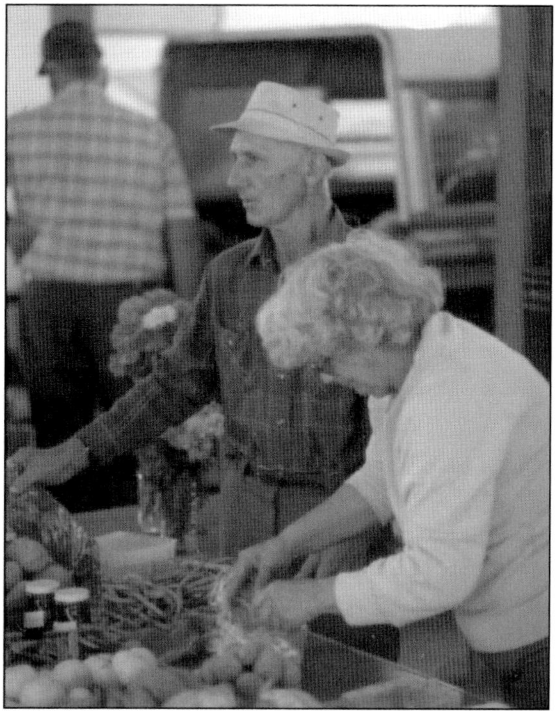

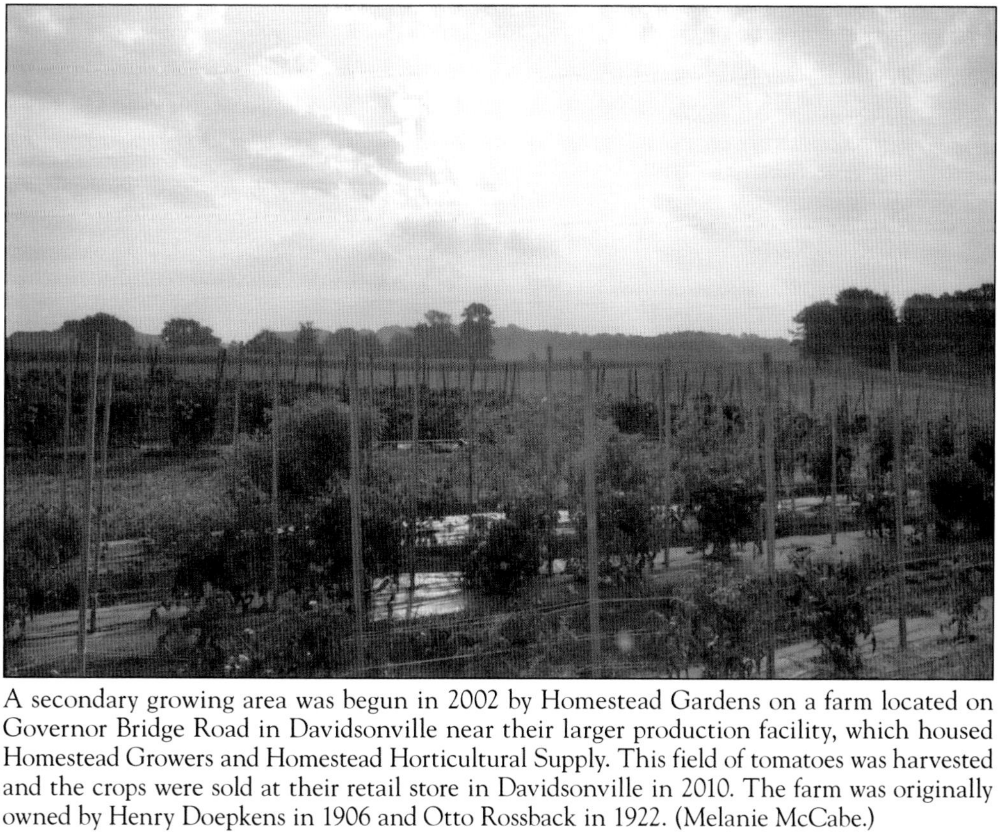

A secondary growing area was begun in 2002 by Homestead Gardens on a farm located on Governor Bridge Road in Davidsonville near their larger production facility, which housed Homestead Growers and Homestead Horticultural Supply. This field of tomatoes was harvested and the crops were sold at their retail store in Davidsonville in 2010. The farm was originally owned by Henry Doepkens in 1906 and Otto Rossback in 1922. (Melanie McCabe.)

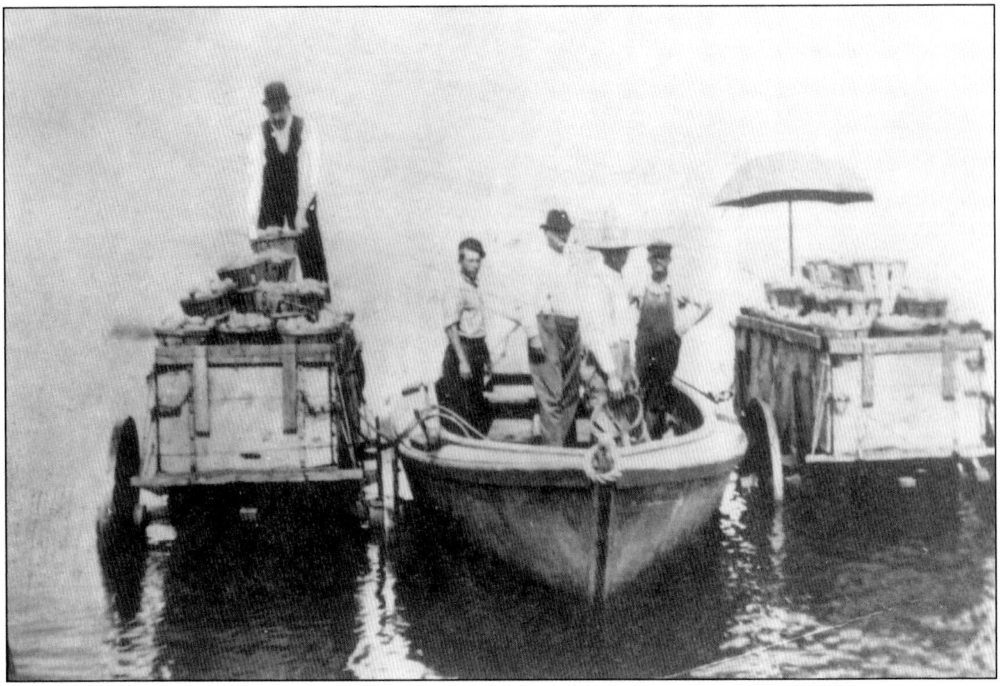

Pictured in 1911 are Alfred Ashbury Stinchcomb's special wagons that were used for loading peaches into a large rowboat in shallow water, as described in Mumford's book on truck cropping; they would later be transferred to a "pungy" boat in deeper water. By 1909, Anne Arundel County was producing over 27,000 bushels of peaches per year that were sold in Baltimore markets. (Kinder.)

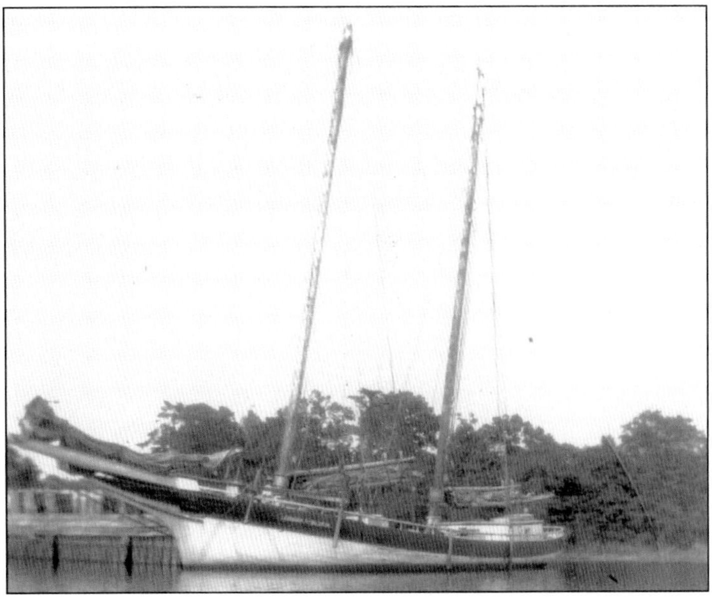

This 1920s photograph shows the *Rhoda Virginia* market boat, which was designed to operate in the shallow waters of Bodkin Creek. Built in 1903 for Henry Alfred Cook and named for his wife, Rhoda Virginia Hancock-Cook, the boat was used to haul produce from the Cook farm and other adjacent farms during the summer and was leased to its captain during the winter months for oystering on the Chesapeake Bay. (Calvert.)

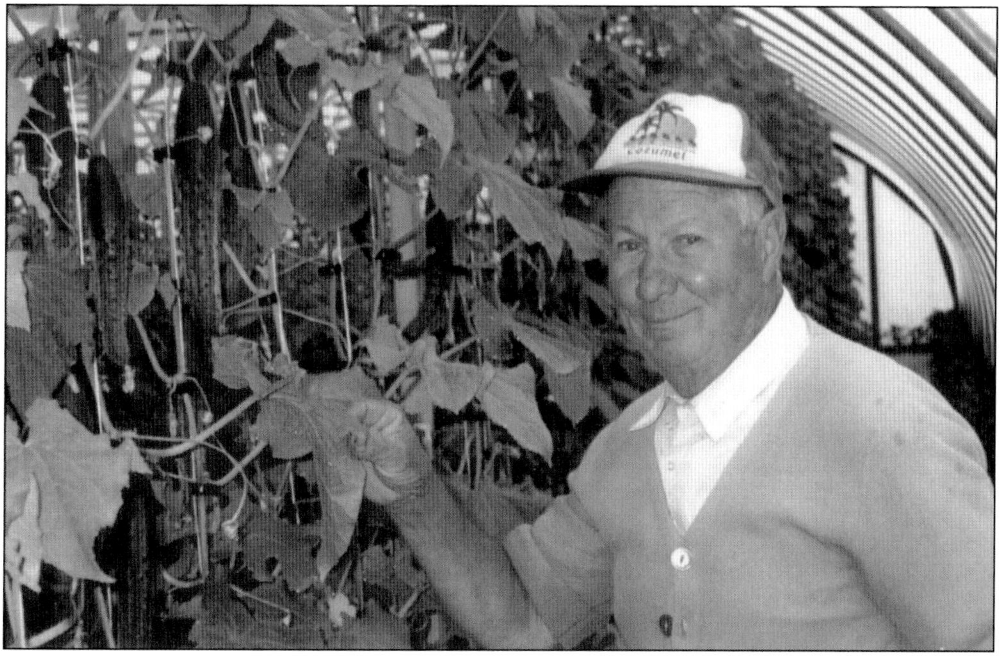

Gordon Wirth shows off his 2009 crop of European cucumbers, which were grown on a trellis on one of the greenhouses on his farm. The cucumbers are harvested when they are 12 inches long and sold at the local farmers' market. Using a greenhouse for growing vegetables can extend the growing season and increase production. (Gordon Wirth.)

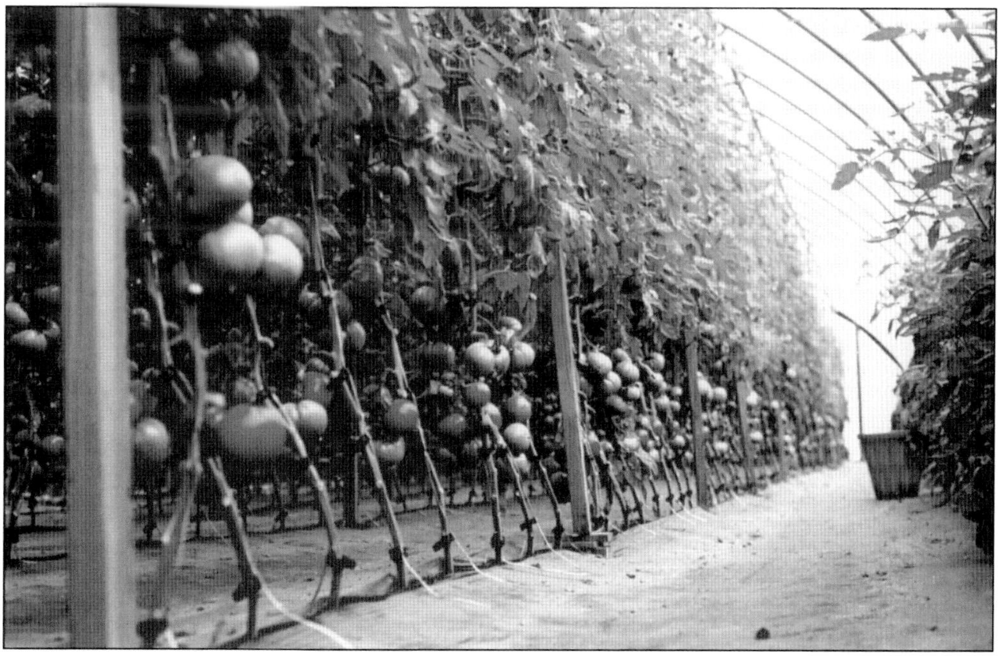

Tomatoes ripen in a greenhouse on the Gordon Wirth farm in Pasadena in 2009. These tomatoes are grown on a trellis system. The fruit is picked once per day and will produce ripe tomatoes for about four months. The fresh-picked fruit is then sold at the local farmers' market on Riva Road near Annapolis. (Gordon Wirth.)

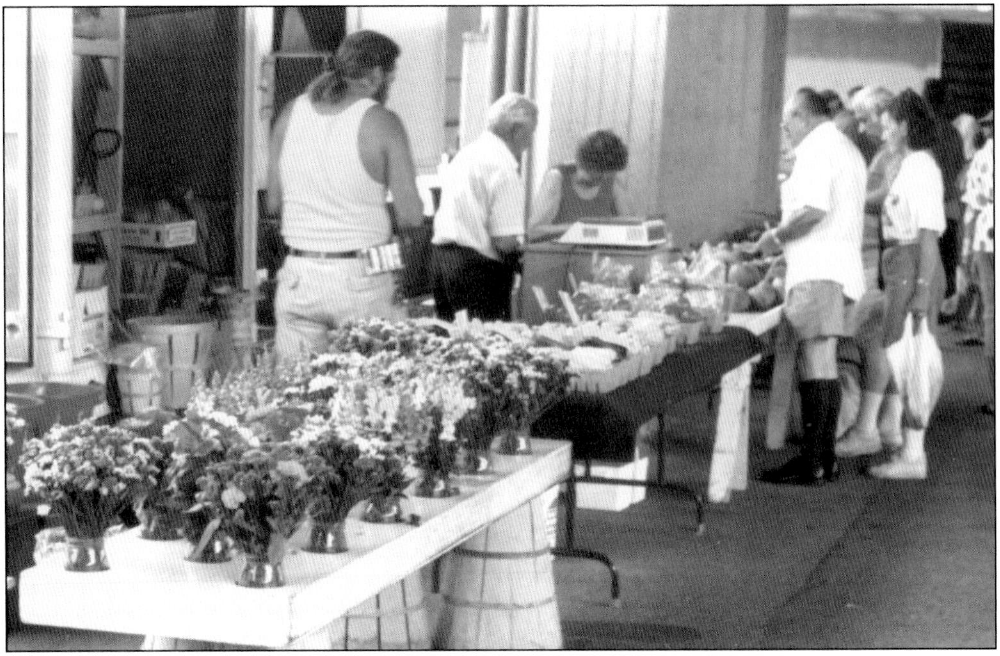

In 1992, the Wirth farm sold its fresh-picked produce at both the Baltimore Farmers' Market and the Riva Road Farmers' Market. In addition to fruits and vegetables from the farm, the Wirths also raise flowers and sell them in fresh-cut bouquets. Flowers such as zinnias, snapdragons, and statice are popular with the customers. (Gordon Wirth.)

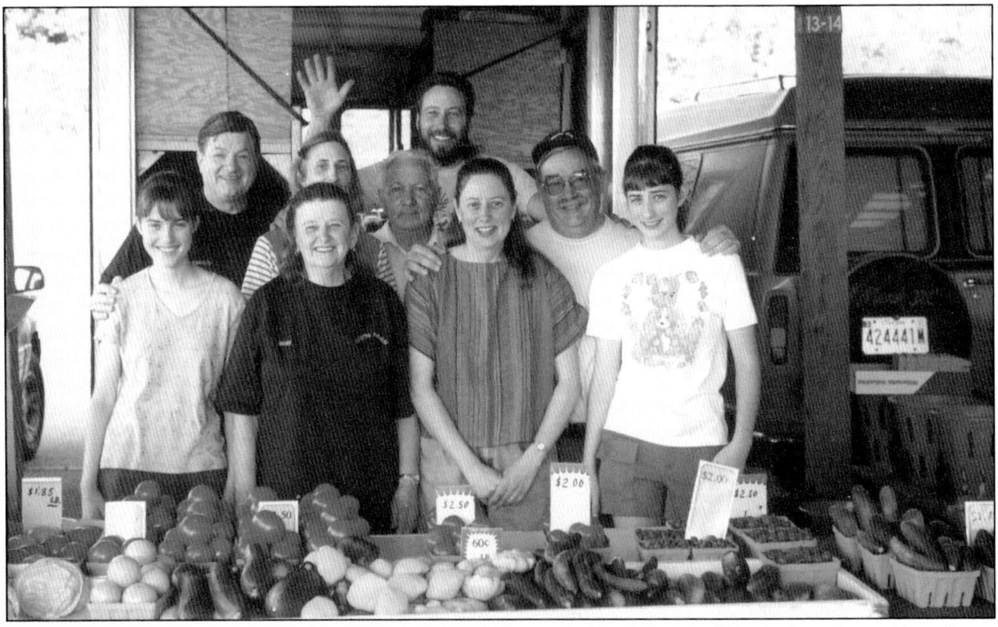

The selling of fresh farm produce is a family affair with the Gordon Wirth family and has been a 30-year tradition at the Riva Road Farmers' Market. Pictured in 2010 are, from left to right, Linda Grimm, Donald and Mary Wirth, Joann Grimm, Gordon and Gary Wirth, Nancy Wirth-Hare, Walter Hare, and Mary Grimm. In 2010, there were eight farmers' markets in the county. (Gordon Wirth.)

Three

Livestock and Dairy Farming

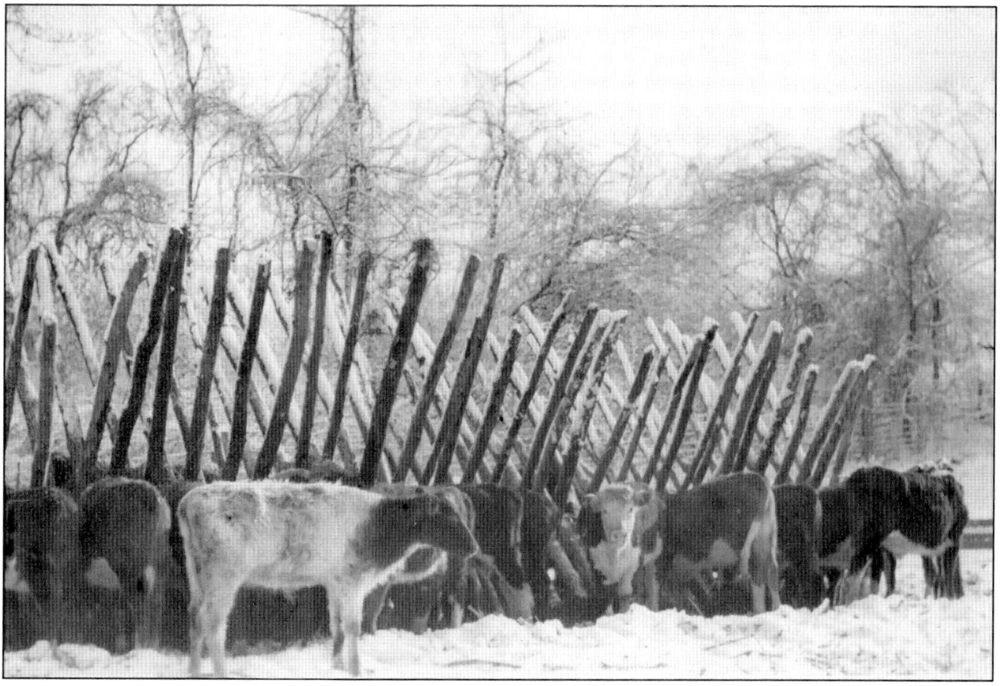

Beef cattle eat hay from a log hayrack on a snowy winter day on Ivy Neck Farm in the early 1900s. Ivy Neck Farm has been in the Colhoun family since the 1680s. In 2010, Allen Colhoun, a ninth-generation farmer on Cumberstone Road near Galesville, produced chicken eggs and grass-fed free-range beef. He sold their Ivy Brand products to several markets in the area. (Colhoun.)

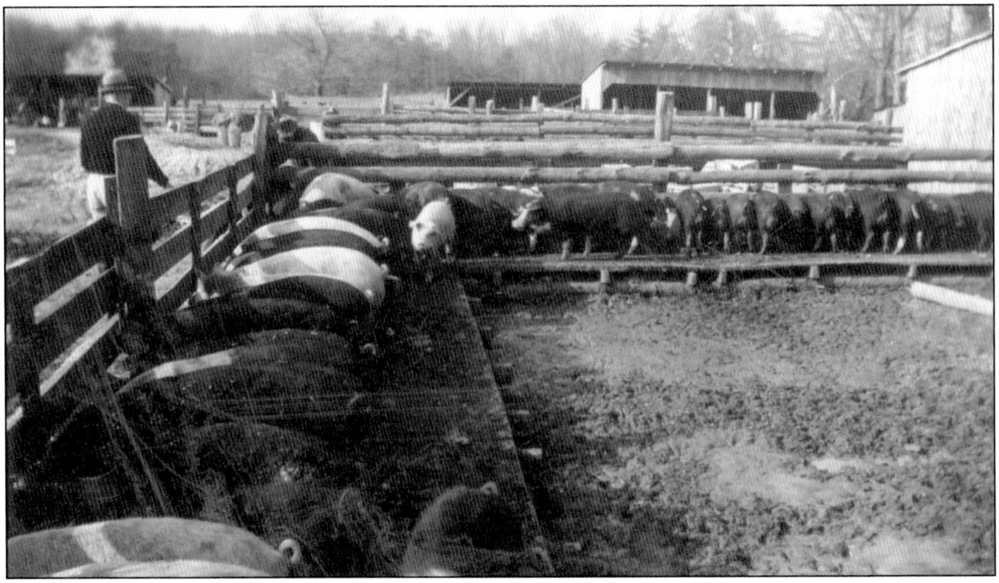

Charles H. Coover Jr. slops hogs on Dixie Farm in Crownsville in the late 1930s. From 1930 through 1952, slop was hauled from the US Naval Academy in Annapolis to feed the hogs on the farm in Crownsville. Charles Coover Sr. purchased the farm in 1929, and by 1943, he had acquired 204 acres. In the 1980s, this part of the farm was taken for the construction of I-97. (Mueller.)

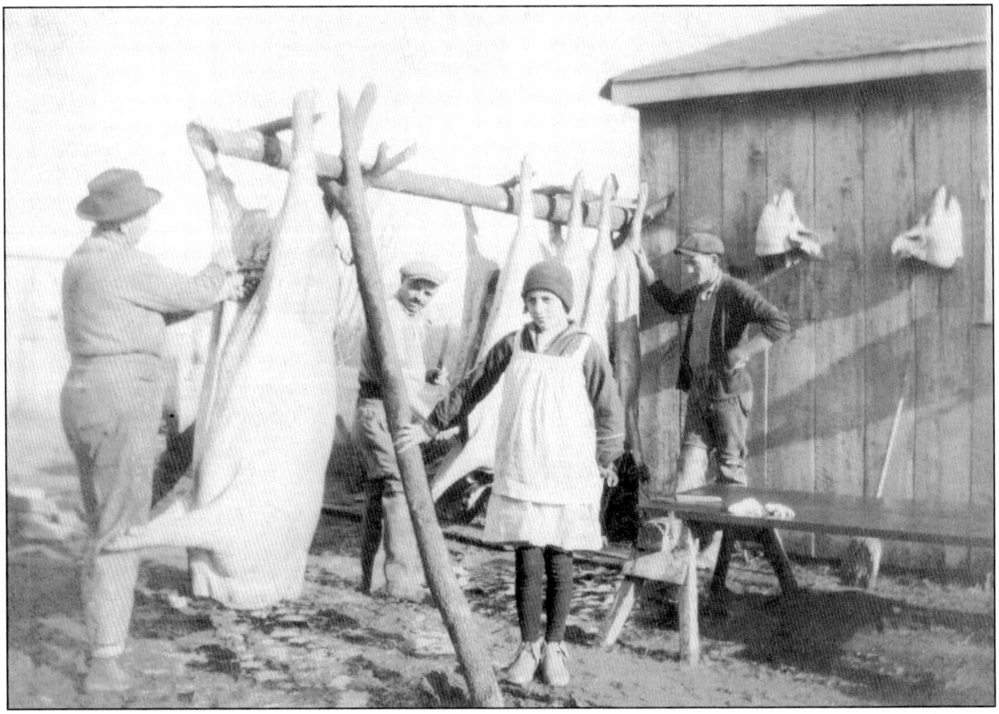

Ernest Mueller Sr. slaughters hogs in the 1920s on his farm on Southaven Road near Annapolis. He would slaughter 10 to 12 hogs per year and make sausage. Ernest Mueller emigrated from his home in Ebendorf, Austria-Hungary, in 1903 and bought a 68-acre farm near Annapolis. (Mueller.)

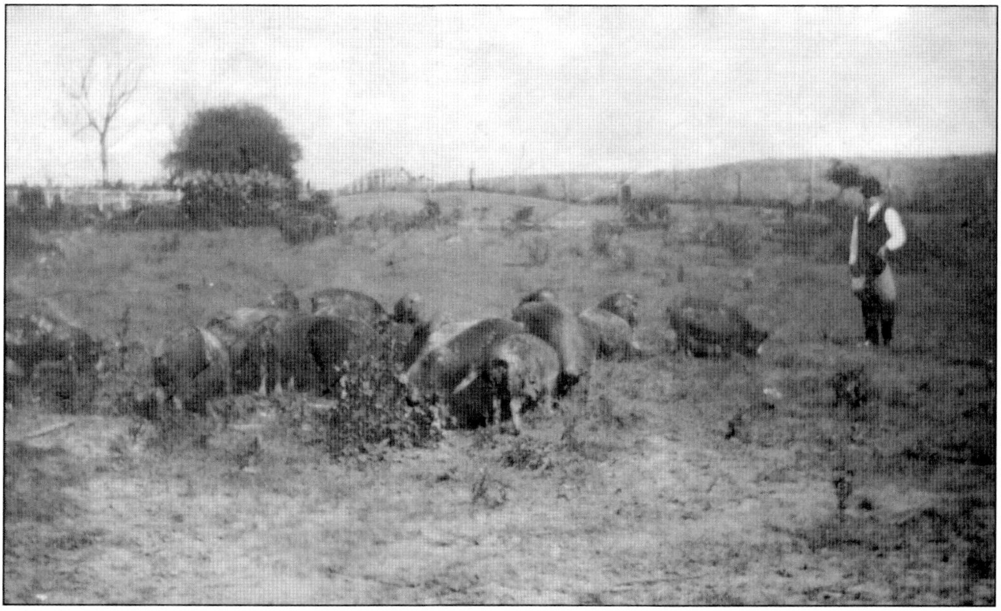

Charles Addison Hodges watches his pastured hogs on his Mount Elm Farm in Davidsonville in 1909. He raised a variety of crops and livestock on the farm. According to the Doepkens book on Middle Plantation, the farm had been in the Hodges family since Charles's grandfather, John T. Hodges, purchased a large tract of land in 1833. (Mary Jay Davis.)

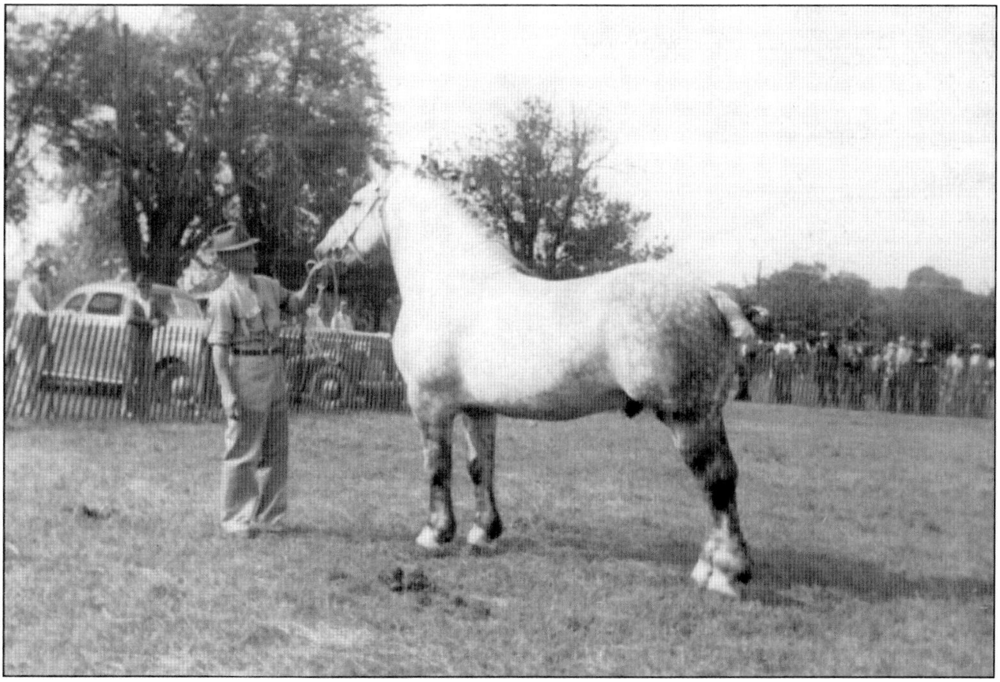

Benjamin Watkins III poses with a Percheron draft horse that he raised on his 120-acre Davidsonville farm around 1940. He also raised a variety of crops and livestock in addition to having a prosperous dairy. His family lived in the Davidsonville area on the Locust Farm for generations. (Catherine Rossback.)

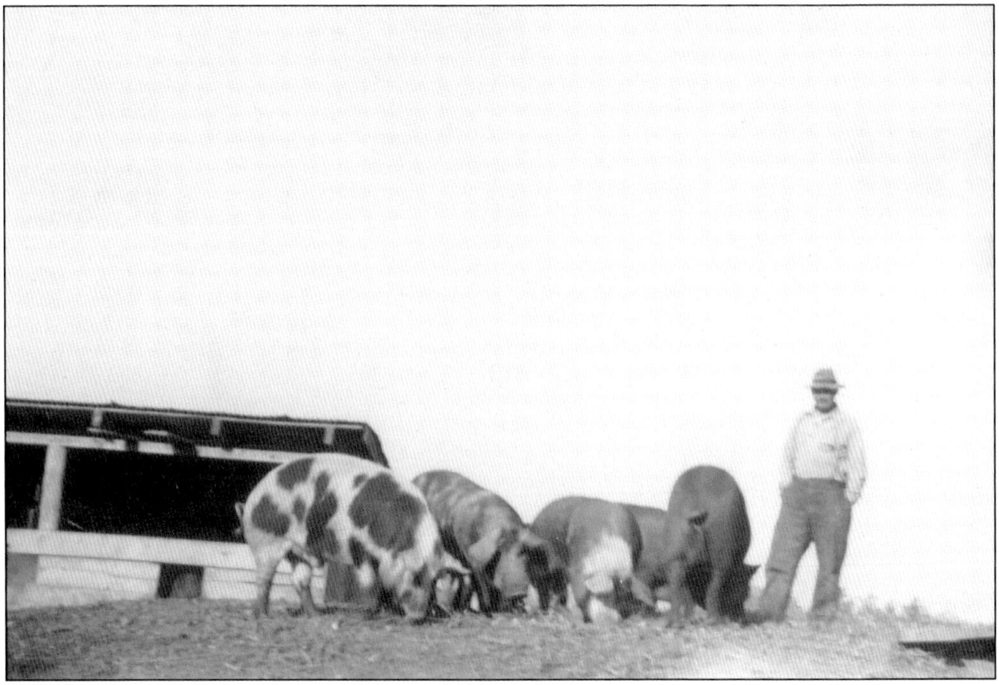

Joseph Bottner is pictured with his Hampshire and Poland China hogs in 1948. He raised 50 to 60 hogs each year on his farm in Gambrills. After butchering the hogs on his farm, he would sell the meat to customers in Annapolis. He emigrated from Germany in 1910 and bought his farm on Davidsonville Road in Gambrills in 1927. (Anderson.)

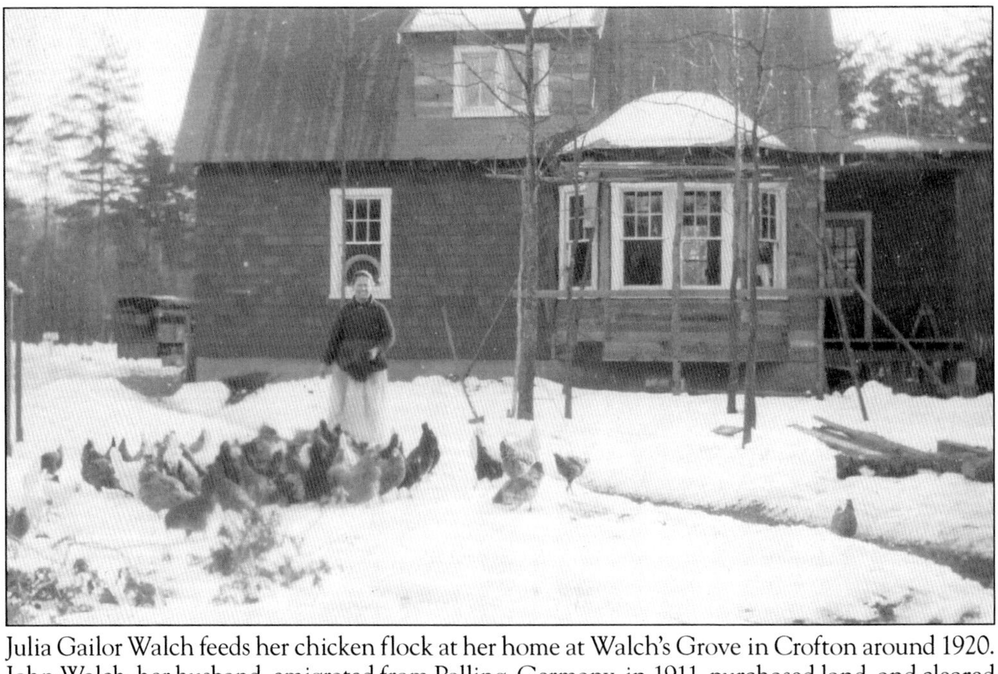

Julia Gailor Walch feeds her chicken flock at her home at Walch's Grove in Crofton around 1920. John Walch, her husband, emigrated from Polling, Germany, in 1911, purchased land, and cleared it for a small farm. (Juliana Walch Ammon.)

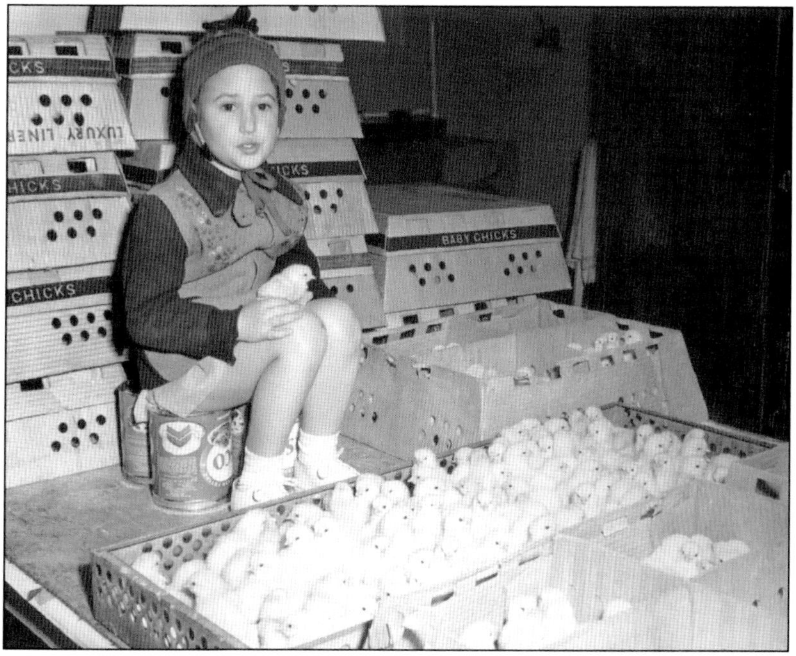

Joann Grimm sits with day-old baby chicks that were hatched in incubators on the Gordon Wirth farm in Pasadena around 1950. They had a large poultry business and would hatch about 1,500 white leghorn chicks per week from January to March. The chicks would be raised and used on their farm for eggs and meat. (Gordon Wirth.)

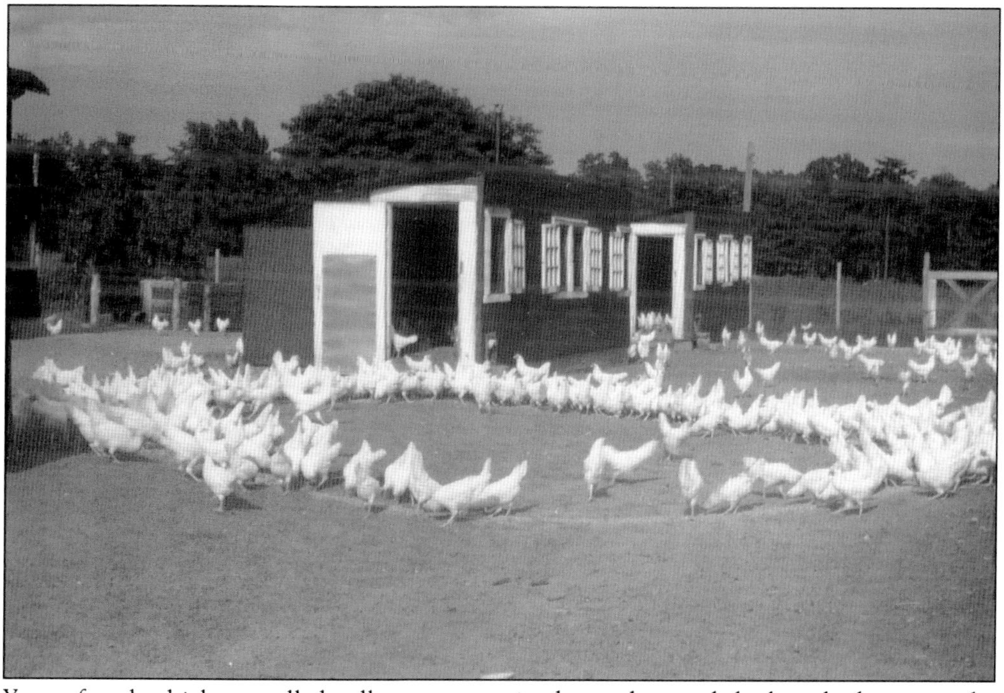

Young female chickens, called pullets, eat corn in the yard around the brooder house on the Wirth farm around 1950. In September, the chickens would be moved to the henhouse for egg production. Hens would be kept for three to four years before being taken to Baltimore and sold as stewing chickens and meat chickens. The Wirths raised chickens from the 1930s until 1962. (Gordon Wirth.)

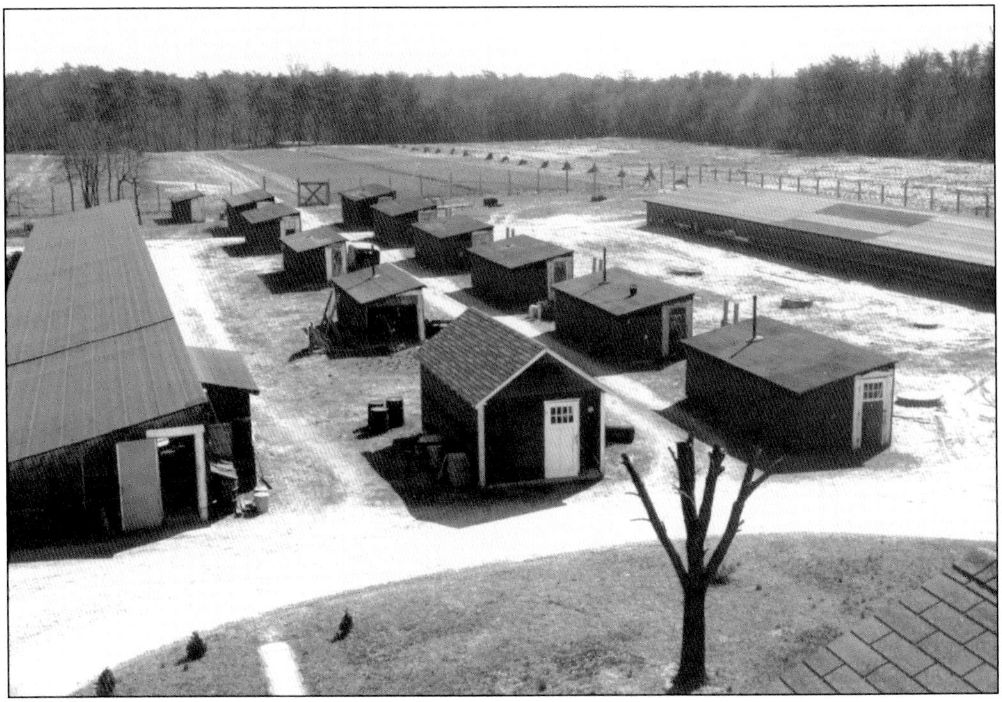

This photograph shows the poultry buildings used on the Wirth farm around 1930. The three long buildings, measuring 20 feet by 210 feet, are the hen houses, which had nine rooms each to house the 6,200 laying hens that produced 3,300 eggs per day. The 12 small buildings in the center are the brooder houses, which were used to raise the chicks to maturity. (Gordon Wirth.)

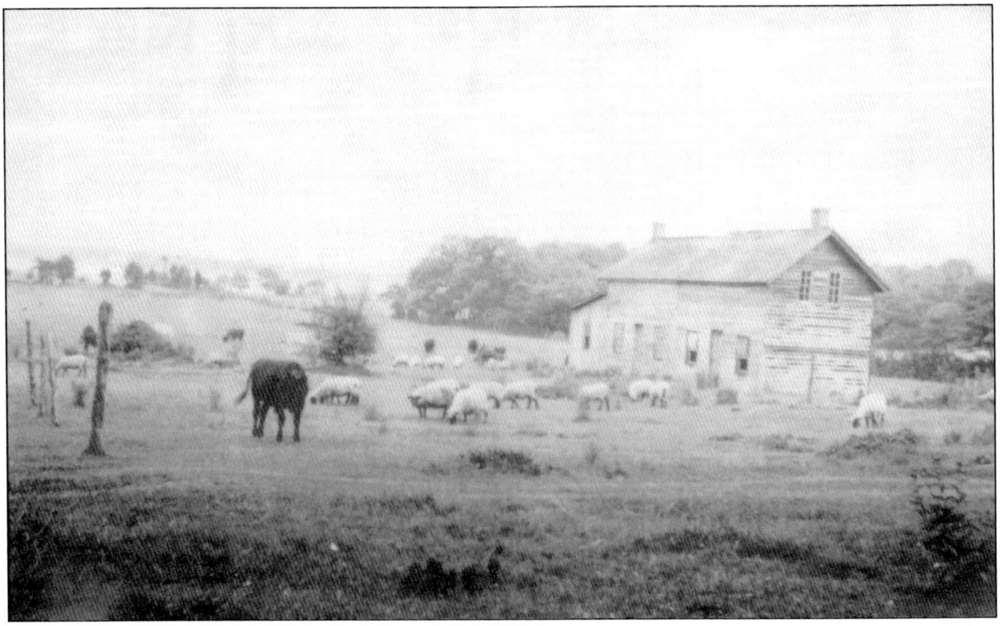

Cattle and sheep graze in the pasture near the tenant house on Ivy Neck Farm around 1940. This house used to be an old slave quarters. (Colhoun.)

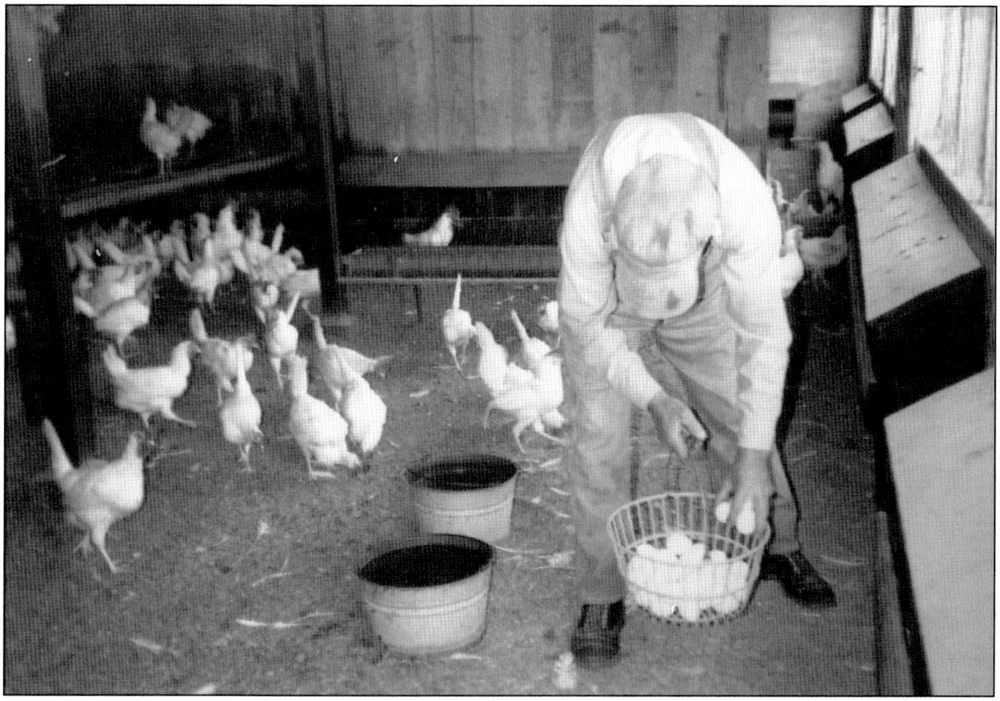

William P. Doepkens collects the eggs from his flock of white leghorn chickens in a wire basket on his farm in 1996. The eggs were usually collected in the late afternoon and processed in the evening. The chickens will lay the eggs in the nest boxes and will lay about one egg per day for about 260 days per year. (FHD.)

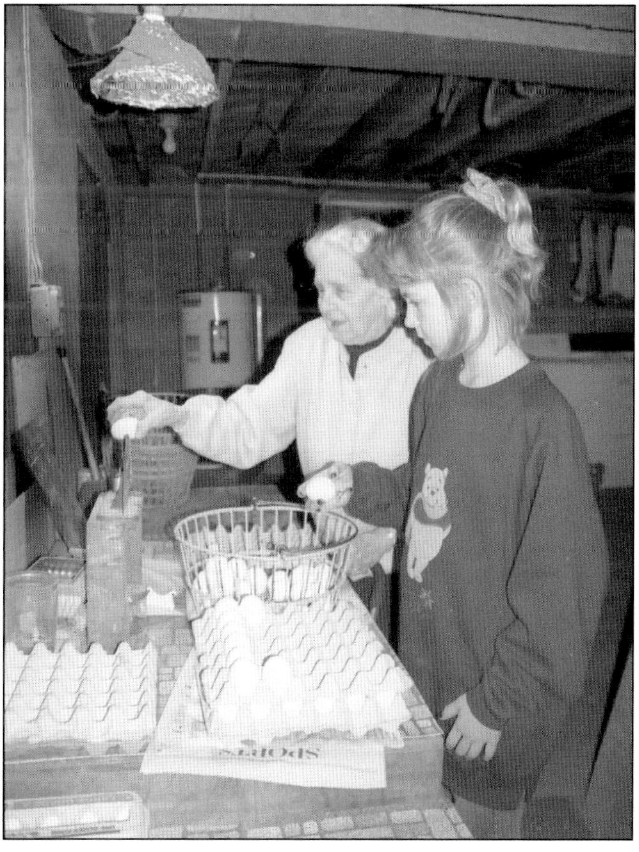

Each day, once the eggs were collected, they were washed and then graded by size. Marjorie D. Doepkens shows her granddaughter, Christina Salyers, how to weigh the eggs using a Jiffy Weigh egg scale around 2000. The eggs were then packed on flats according to size: small, medium, large, extra large, or jumbo. The eggs were sold to customers who came to the Doepkens farm. (FHD.)

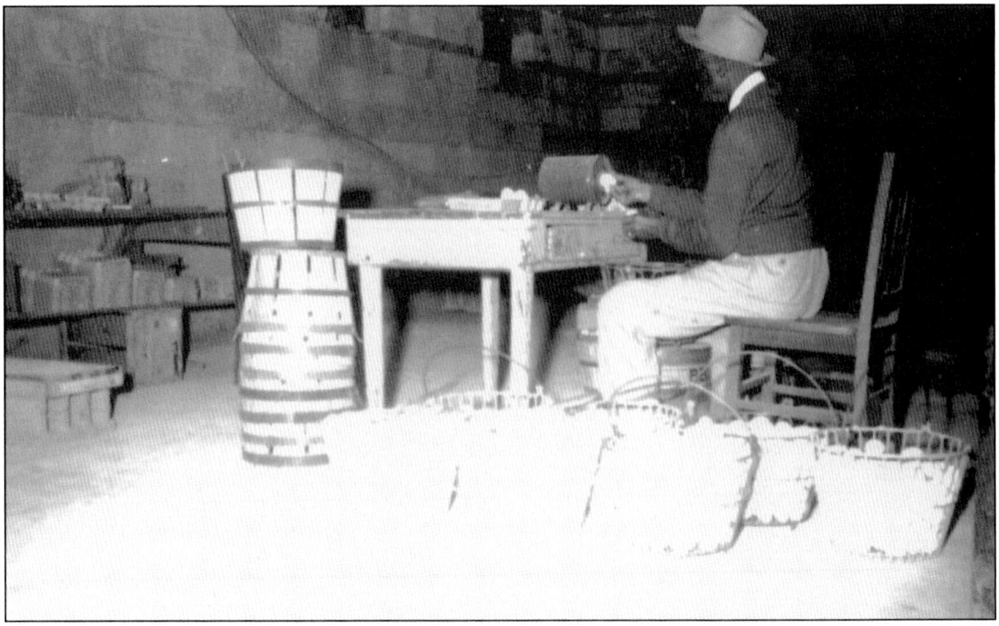

To ensure top-quality eggs, Walter Thompson candles the eggs produced on the Caldwell McMillan farm in Crownsville around 1950. The candle shines a light through the egg, which will enable the farmer to see if the egg is cracked or has blood spots. (Caldwell McMillan.)

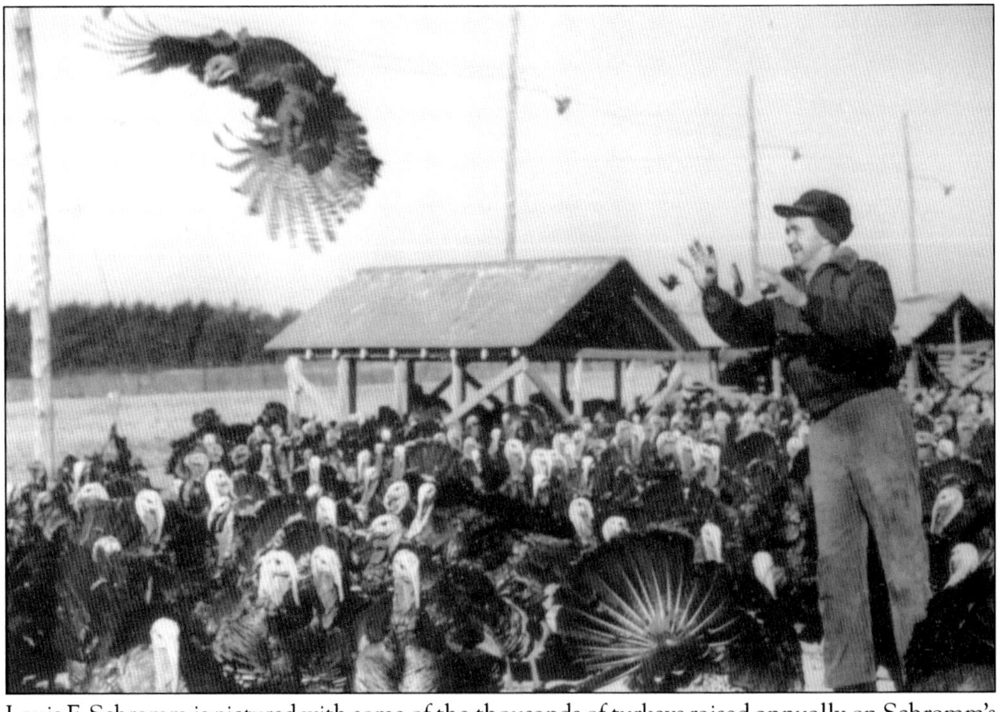

Louis F. Schramm is pictured with some of the thousands of turkeys raised annually on Schramm's farm in Pasadena in 1955. The shelters for the turkeys can be seen in the background. The Schramms slaughtered and processed their turkeys on the farm to prepare them for sale to customers. (Sun.)

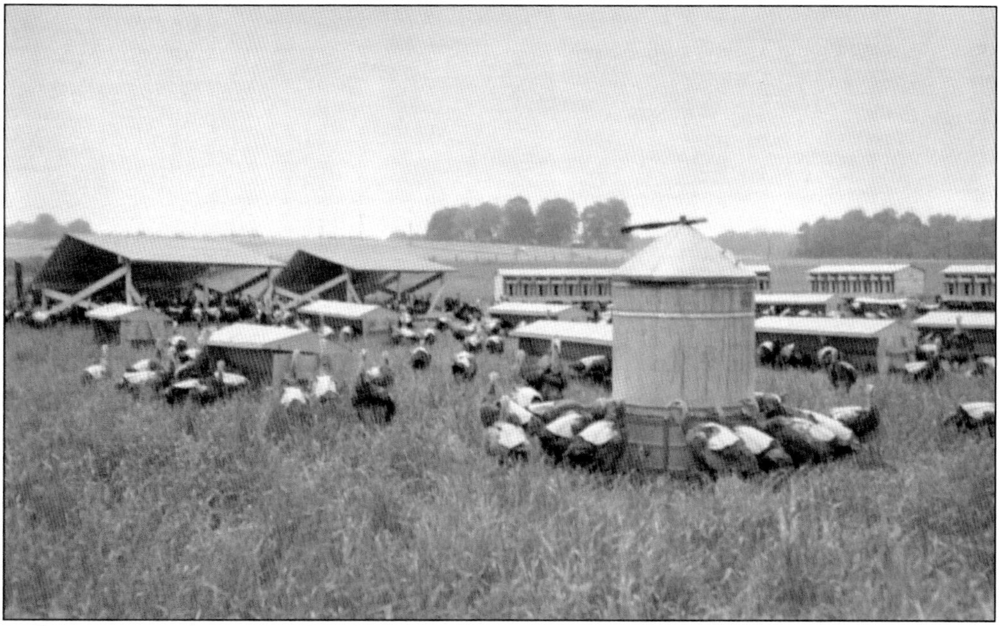

Turkeys graze on the Taylored Turkey Farm on Route 3 in Gambrills around 1950. Presley Taylor Jr. operated the turkey hatchery and breeding farm. He raised broad-breasted bronze turkeys and sold both young turkeys and processed turkeys to wholesale and retail markets. (Colhoun.)

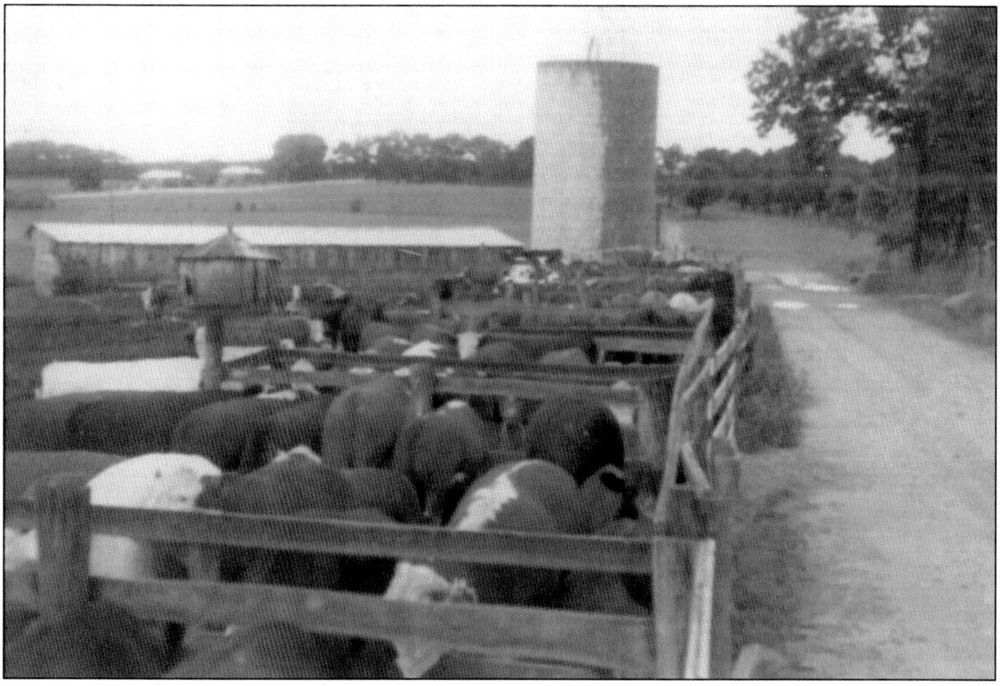

Beef cattle stand in the feedlot on the Albert and August Kinder farm in Millersville in the 1960s. Cattle were purchased at 500 pounds and then fed and raised until they reached 1,200 pounds; then they were sold to Baltimore Union Stockyards. A special mash from the Monumental Distillery was fed to the cattle to achieve rapid weight gain. The farm was sold in the late 1970s. (Kinder.)

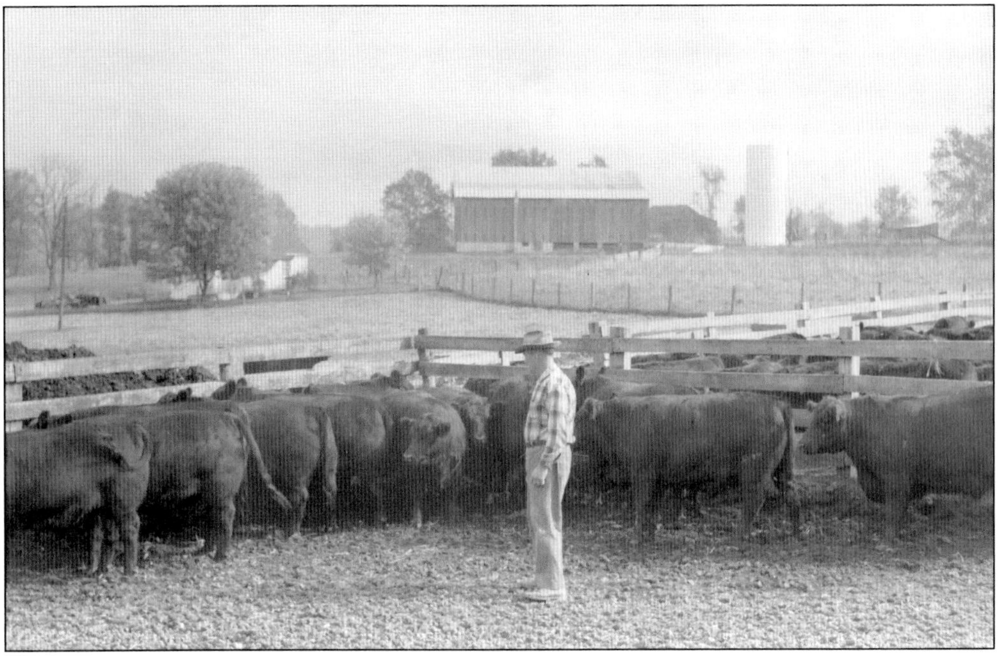

In 1937, Maxwell H. Covington Sr. moved to the Davidsonville area and purchased a 240-acre farm from Wax Townsend. In this c. 1950 photograph, Max inspects his herd of 200 Black Angus beef cattle. Soybeans, corn, and hay were raised on the farm in addition to 35 acres of tobacco. The family sold the herd around 2000. (Covington.)

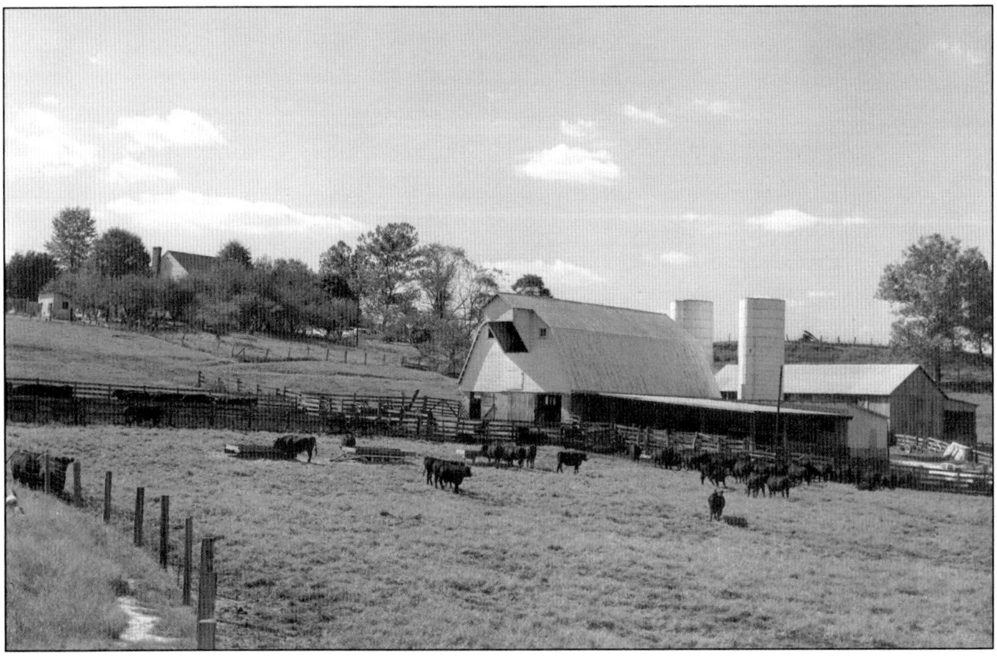

Pictured is the beef cattle barn on the Covington farm with the silos for storing feed for the animals around 1950. Many of the silos built in the Davidsonville area were constructed by Max Covington Sr., who owned a franchise for the Marietta Silo Company. Hay was stored in the top of the barn. A cattle loafing shed can be seen here (attached to the barn). (Covington.)

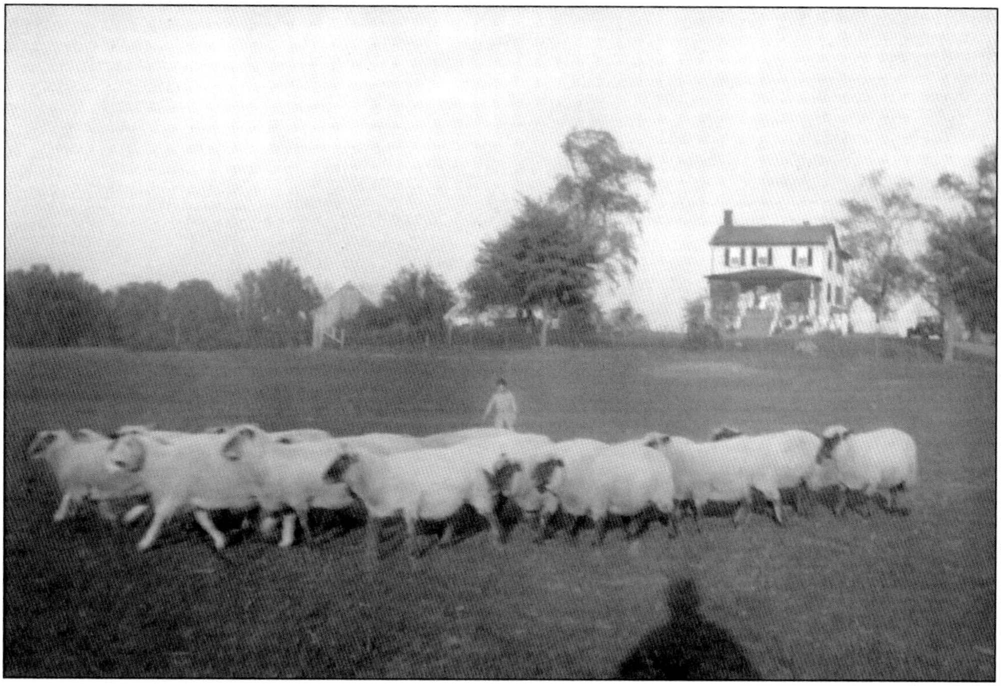

Sheep were grown on many farms. Oscar F. Grimes Jr. (eight years old) stands behind the flock of Hampshire Cross sheep in 1932. These sheep were raised for wool, and the spring lambs were sold to merchants in Annapolis. The Grimes family's Davidsonville farmstead is seen in the background. His father, Oscar F. Grimes Sr., purchased the farm in 1926 from Emory Bassford. (Grimes.)

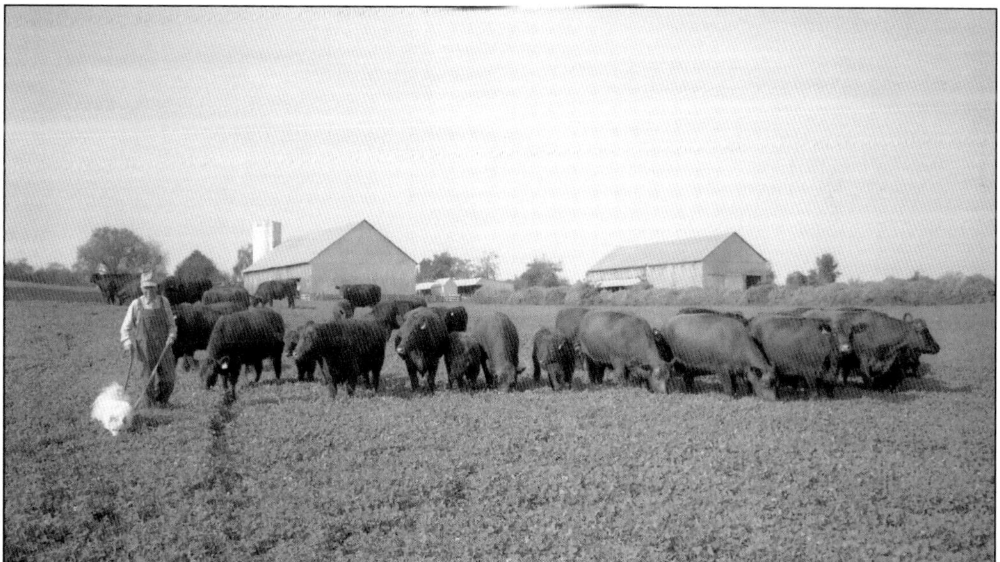

William P. Doepkens and his dog, Willie, are seen in the pasture with this herd of Red Angus beef cattle in 1987. About 24 head were usually kept on the farm and raised for meat for the family and for sale. The family farmstead, comprised of the home, cattle barn, and tobacco barn, appears in the background. William's father, Henry Doepkens, purchased the farm on Davidsonville Road in 1922. (FHD.)

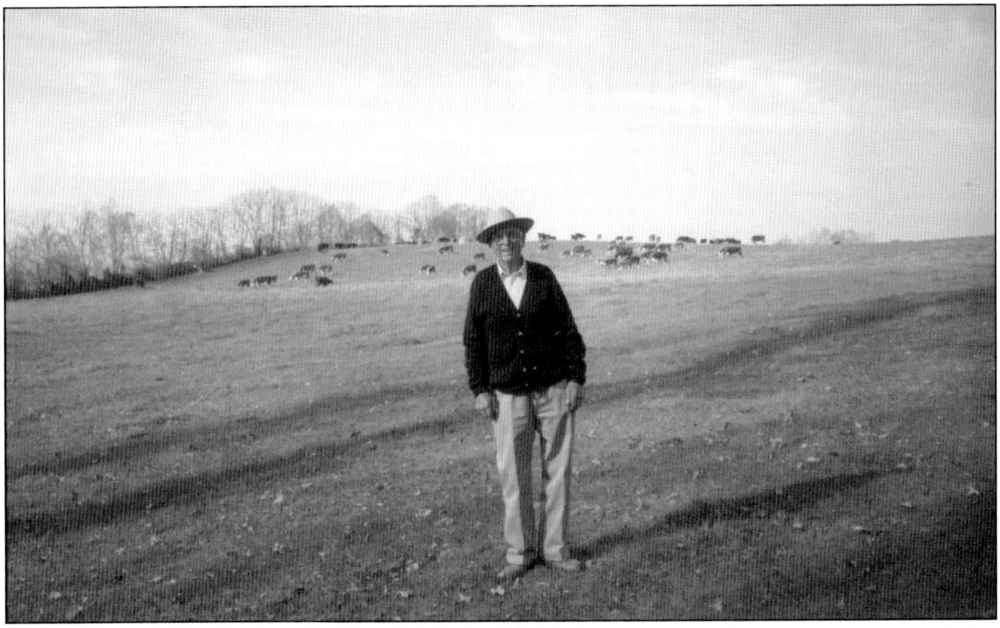

John Robert Phibbons is pictured on his farm, Hill Valley, on Bayard Road in Lothian in 1994. John was a cattle farmer for many years. He and his wife, Jessie, also hauled livestock to and from market for farmers in southern Maryland. John was proud of the registered Polled Hereford cattle he raised, which were highly prized by cattle buyers. (Doris S. Phibbons.)

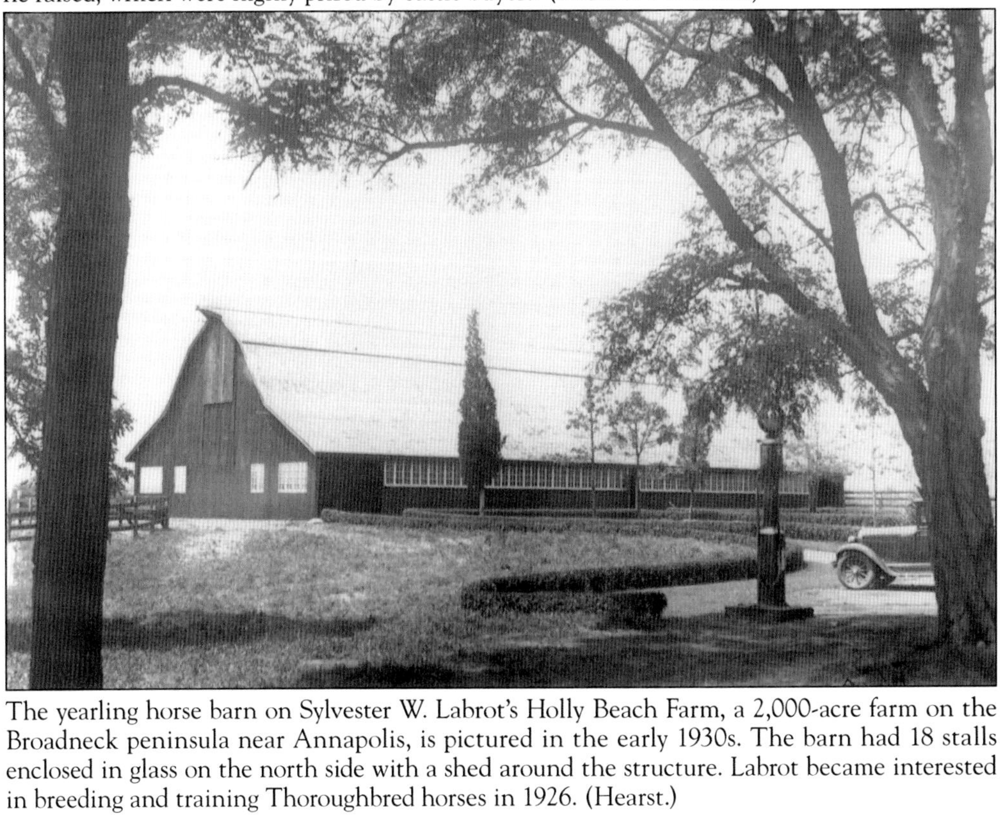

The yearling horse barn on Sylvester W. Labrot's Holly Beach Farm, a 2,000-acre farm on the Broadneck peninsula near Annapolis, is pictured in the early 1930s. The barn had 18 stalls enclosed in glass on the north side with a shed around the structure. Labrot became interested in breeding and training Thoroughbred horses in 1926. (Hearst.)

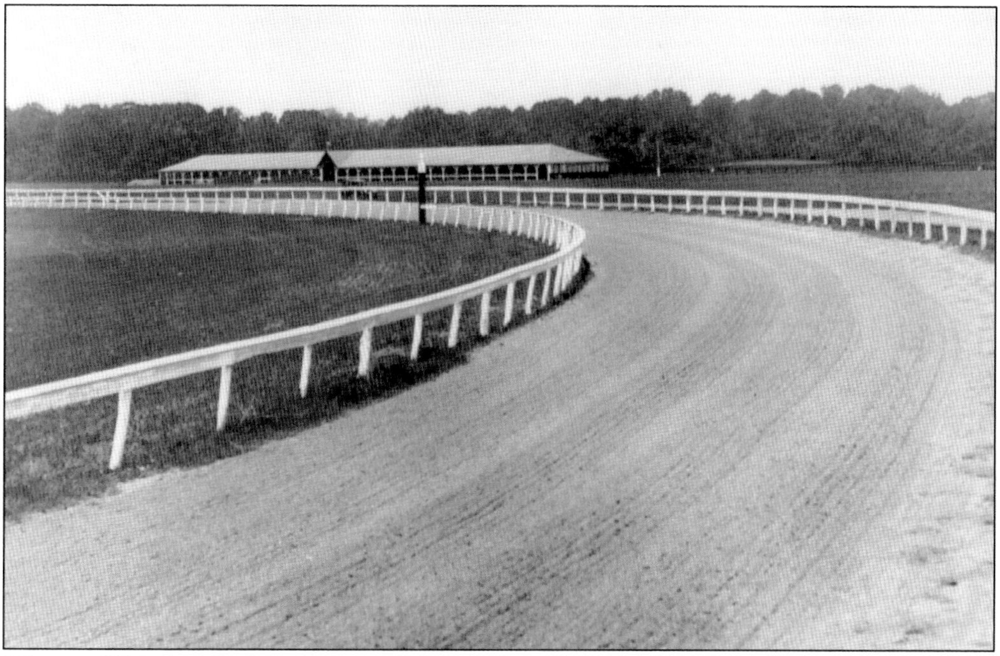

Sylvester W. Labrot built a private mile-long racetrack on his farm around 1929; the track was active until after World War II. In this c. 1930 photograph, his training stable can be seen in the background. The first Anne Arundel County Fair was held on this racetrack in 1953. This property is now part of the Revell Downs Development. (Hearst.)

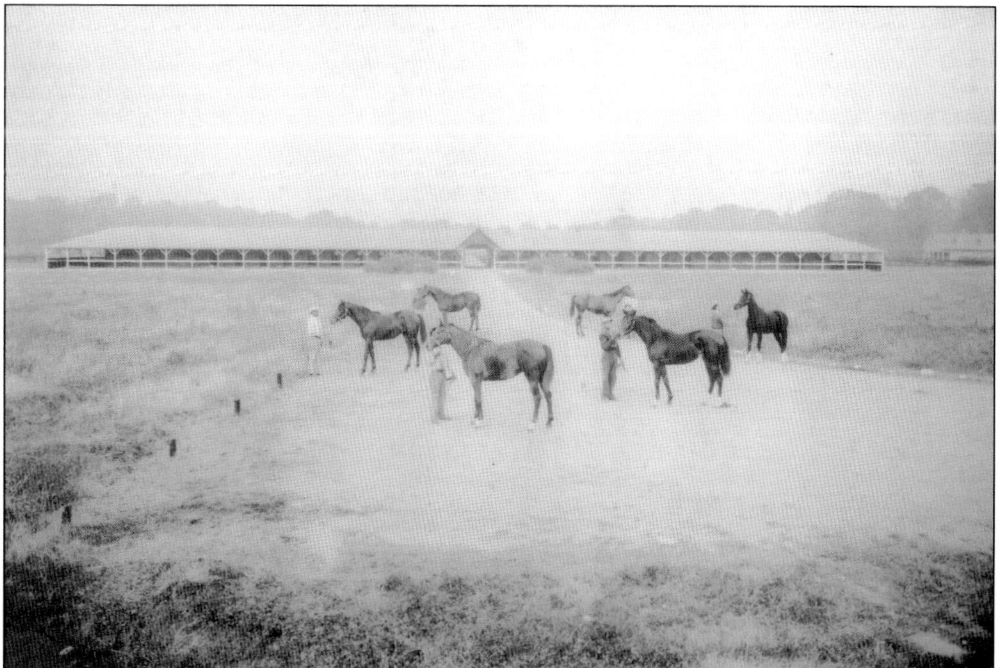

Five Thoroughbred horses and handlers pose in front of the big training barn on Holly Beach Farm around 1932. By the early 1930s, the farm became one of the largest Thoroughbred breeding and training establishments in the United States. (MSA.)

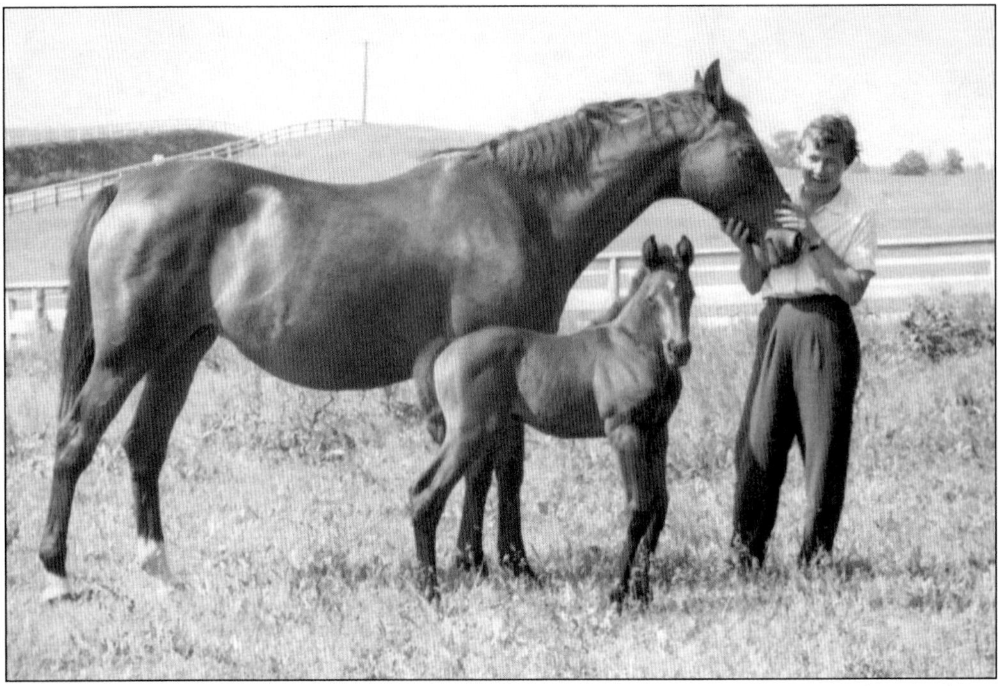

Genie Tucker is shown with one of her horses and its colt on her farm in Davidsonville around 1940. In addition to raising and showing horses, she worked alongside her husband, William O. Tucker Jr., to support the family's farm and lumber business. (Chase.)

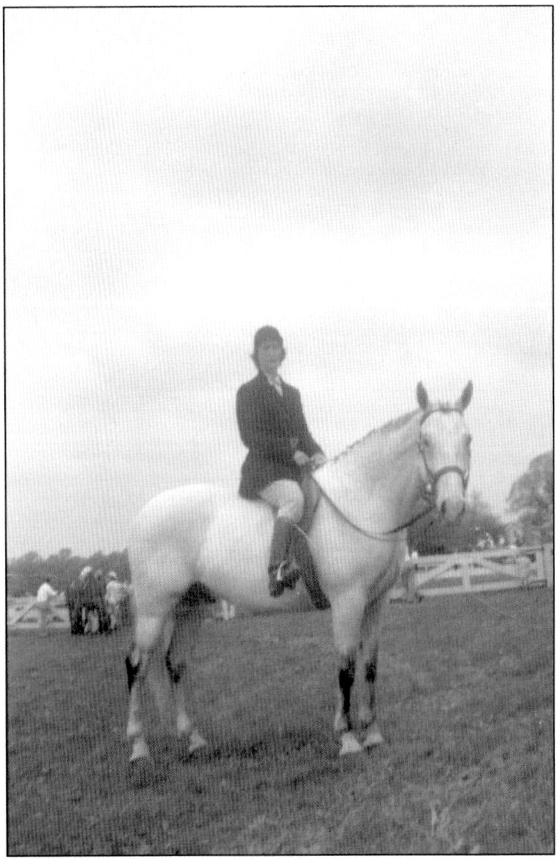

In 2007, Marilyn Hugg was inducted into the Maryland Horse Show Association's Hall of Fame. Here Marilyn rides David Gray, one of her champion horses, around 1972. In addition to participating in horse shows and training her own horses and ponies in the hunter division, she also taught lessons for many children. In 1987, she retired after 42 years of showing. (Charles Hugg.)

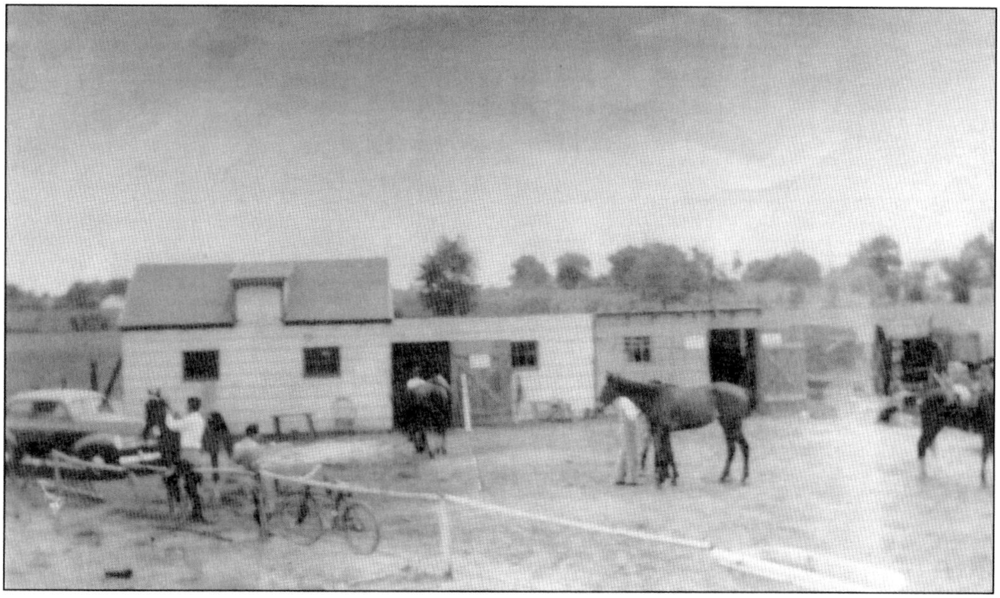

Edward Hugg moved in 1940 to Annapolis, where he and his son, Charles, built these barns on the property. Due to the farm's proximity to the Naval Academy, midshipmen would come over to rent horses for 75¢ an hour to ride. In 1941, the Huggs had four acres, which included a barn and pasture on the property. (Charles Hugg.)

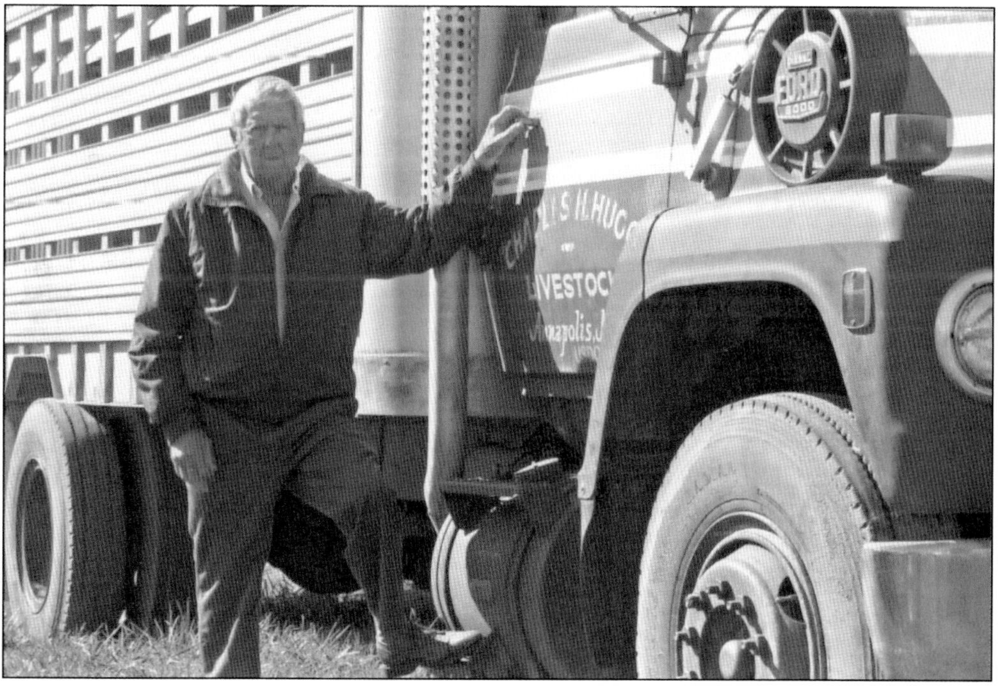

For over 64 years, Charles Hugg hauled livestock for farmers in Anne Arundel County. As an Army veteran, in 1946, he purchased his first truck from Fort Meade and then built the body for the truck so he could haul animals. He began buying and selling horses and mules and eventually expanded to cattle and other livestock. This photograph was taken in 2010. (Charles Hugg.)

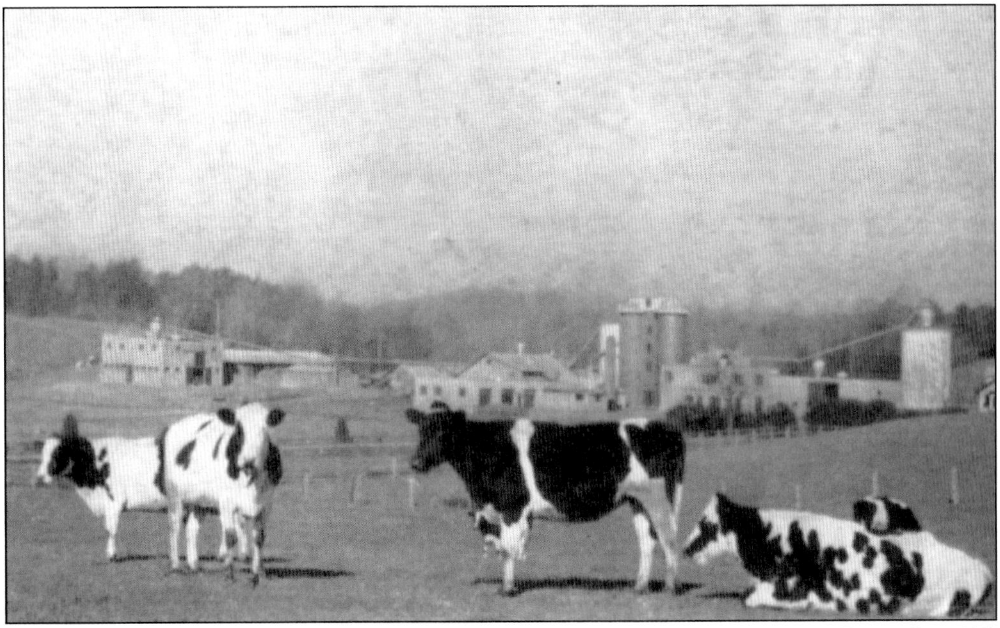

In 1915, the 368-acre Java Dairy Farm was started by Robert L. Forrest in West River. By 1935, a herd of 150 Holstein dairy cattle produced milk for over 300 customers, including the Naval Academy and the Annapolis Hospital. The milk was either pasteurized or certified disease free. In 1940, Forrest sold the herd due to lack of labor before World War II. In 1962, the Smithsonian acquired the property. (SERC.)

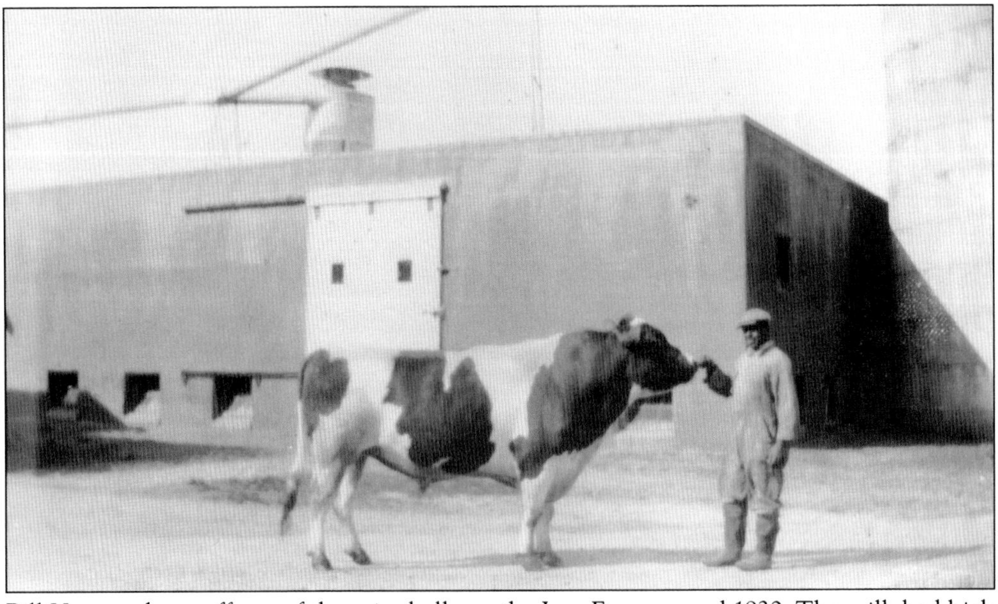

Bill Henson shows off one of the prize bulls on the Java Farm around 1930. The milk had high concentrations of vitamins due to the cultivation of special feeds and a special method of drying hay using a hay dehydrator, the first of its kind in Maryland. The dryer dried the hay more rapidly and retained more nutrients. Then the hay was stored in a silo until it was fed to livestock. (SERC.)

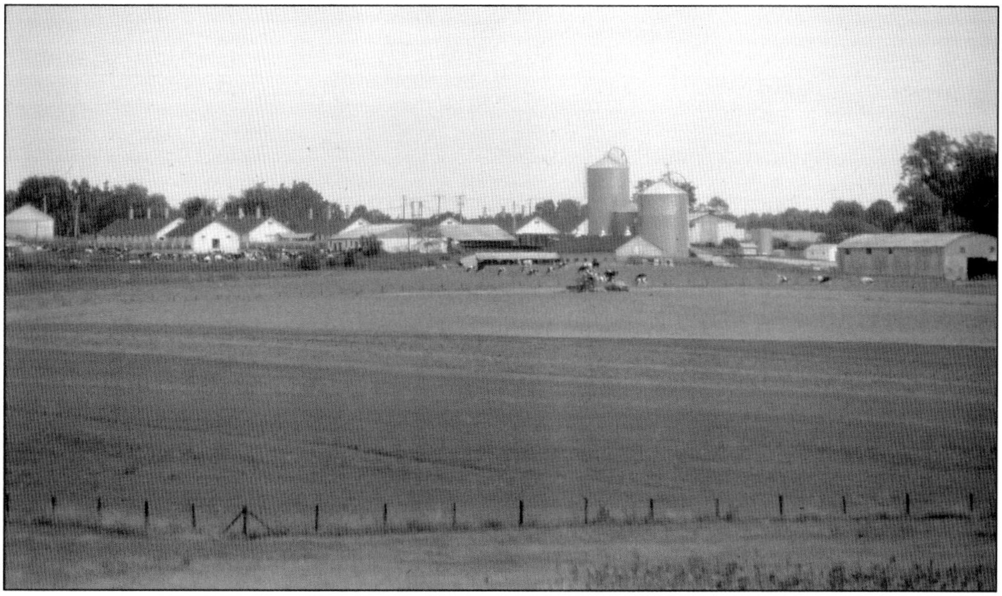

This August 1991 view shows the US Naval Academy Dairy Farm in Gambrills. In 1914, the US Naval Academy purchased the 857-acre Hammond Manor farm. The dairy farm was established in 1918 and was the only US Department of Defense facility of its kind, designed to provide a safe and sanitary milk supply to the midshipmen at the US Naval Academy in Annapolis. (AASCD.)

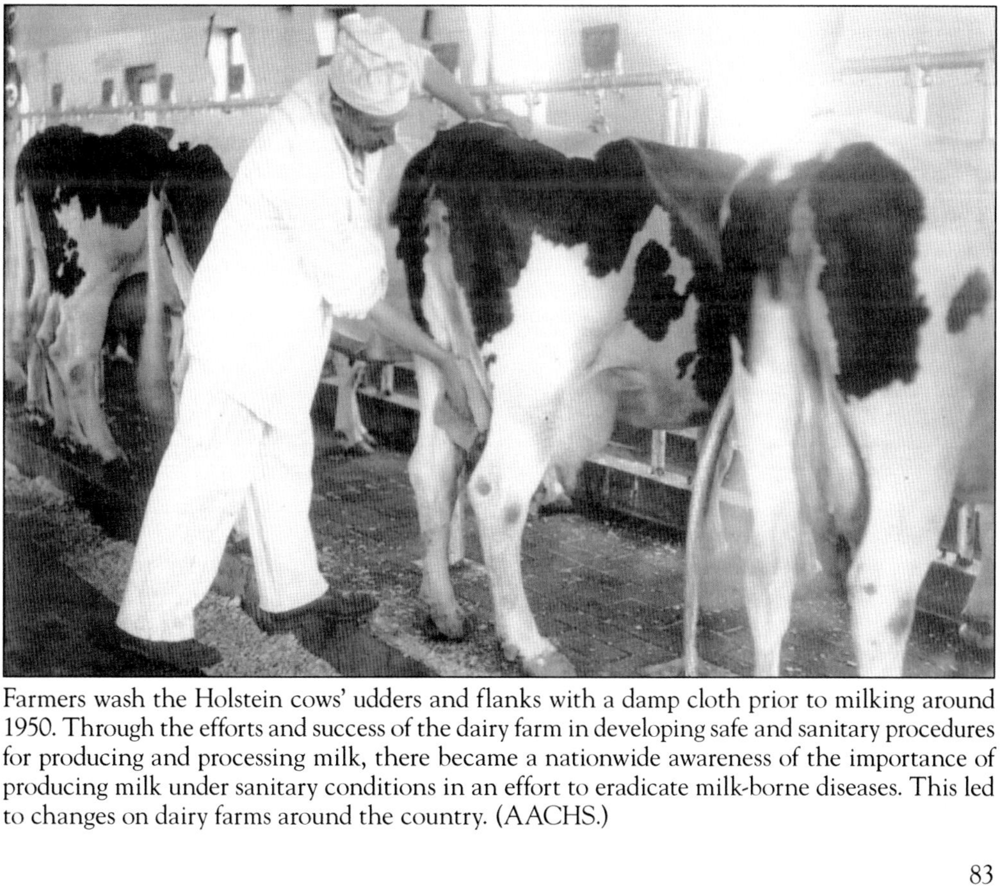

Farmers wash the Holstein cows' udders and flanks with a damp cloth prior to milking around 1950. Through the efforts and success of the dairy farm in developing safe and sanitary procedures for producing and processing milk, there became a nationwide awareness of the importance of producing milk under sanitary conditions in an effort to eradicate milk-borne diseases. This led to changes on dairy farms around the country. (AACHS.)

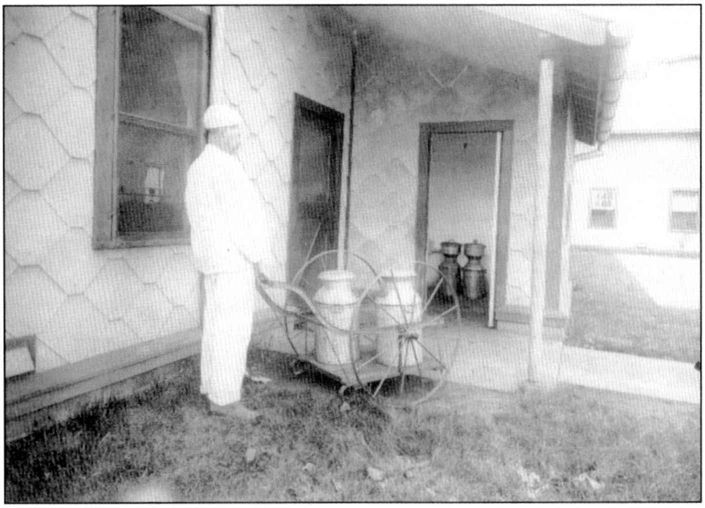

After the cattle have been milked, the milk is transferred from the weighing room to the milk house in large stainless steel milk cans around 1950. There were 200 head of milk cows in the herd in 1917. On average, a single dairy cow can produce up to eight gallons of milk, consume about 80 pounds of feed, and drink 30 to 40 gallons of water per day. (AACHS.)

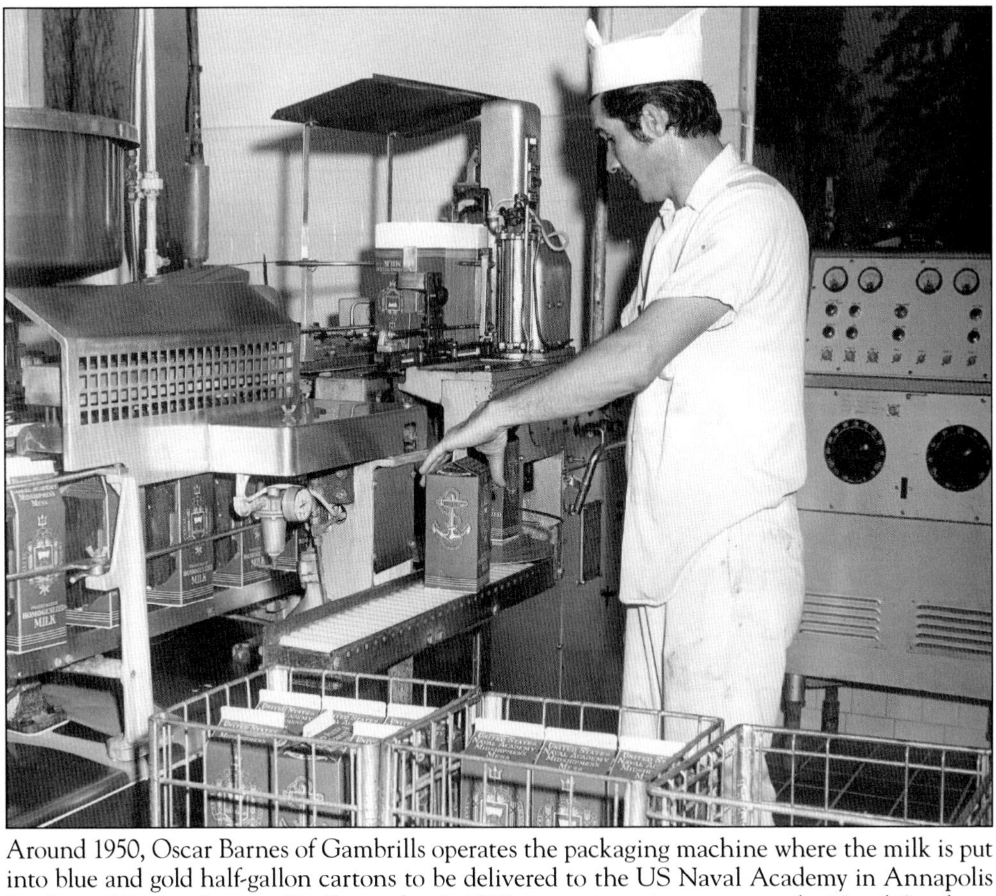

Around 1950, Oscar Barnes of Gambrills operates the packaging machine where the milk is put into blue and gold half-gallon cartons to be delivered to the US Naval Academy in Annapolis for the midshipmen's mess. On September 2, 1998, in a cost-saving measure, the Naval Academy Dairy, the last operating dairy farm in Anne Arundel County, closed its doors and sold its herd of 140 cattle. (AACHS.)

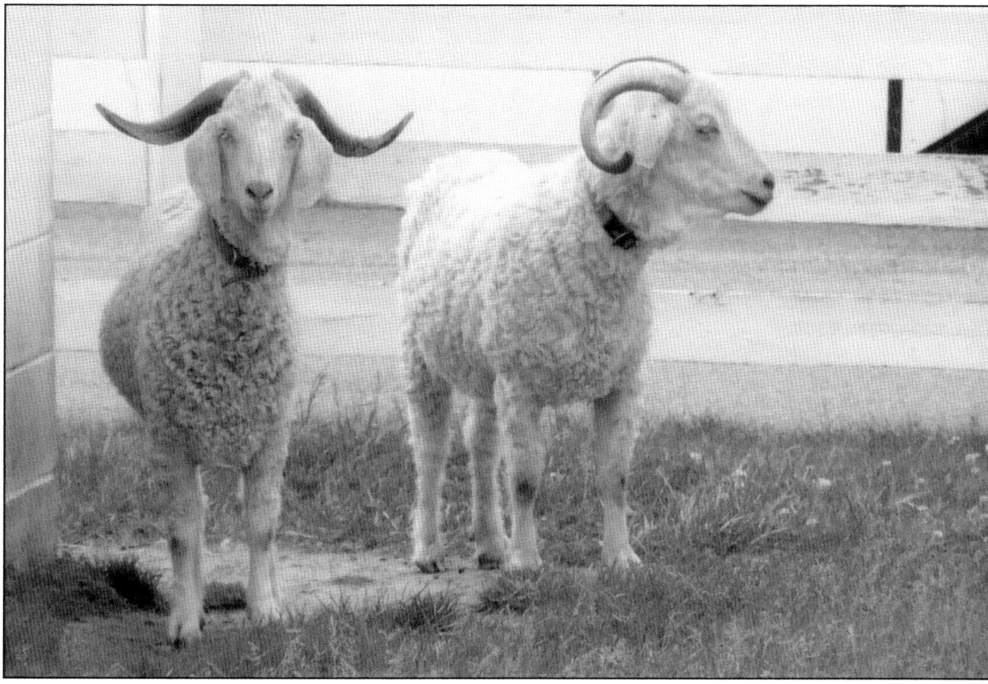

During the operation of the US Naval Academy Dairy Farm, the Naval Academy mascot, "Bill the Goat," was raised on the farm. In 2010, the Angora goats used as the mascots were cared for by Team Bill, a group of midshipmen from the US Naval Academy. Maryland Sunrise Farm, LLC, now operates the farm as part of the Anne Arundel County Department of Recreation and Parks. (FHD.)

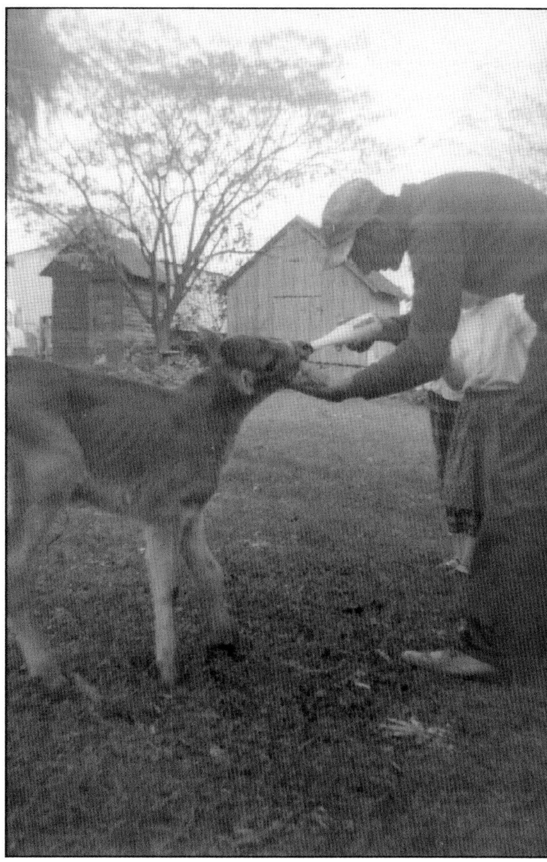

Oscar F. Grimes Jr. bottle-feeds a Guernsey calf that is being raised for veal on his farm in 1955. He usually purchased Holstein calves from other dairy farms in the area and then raised them for veal calves. The calves would be fed for about 18 weeks until they were about 450 pounds and were then sold for meat. The smoke- and meat-houses can be seen in the background. (Grimes.)

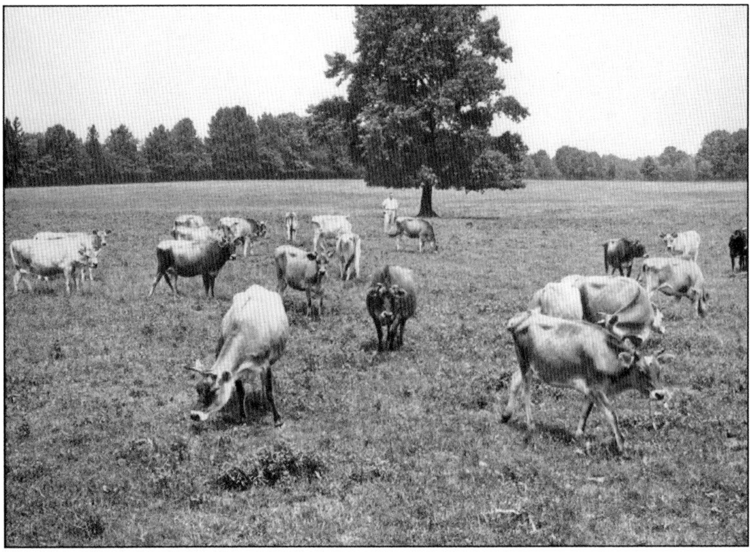

In 1907, Sylvester W. Labrot purchased a 2,000-acre farm on the Broadneck peninsula and quickly developed it into a showplace for horse breeding and dairy cattle. His son, William, operated a large dairy with a herd of Jersey cattle, seen here around 1930 on the Holly Beach Dairy Farm with the dairy manager, Robert Tilghman. (MSA.)

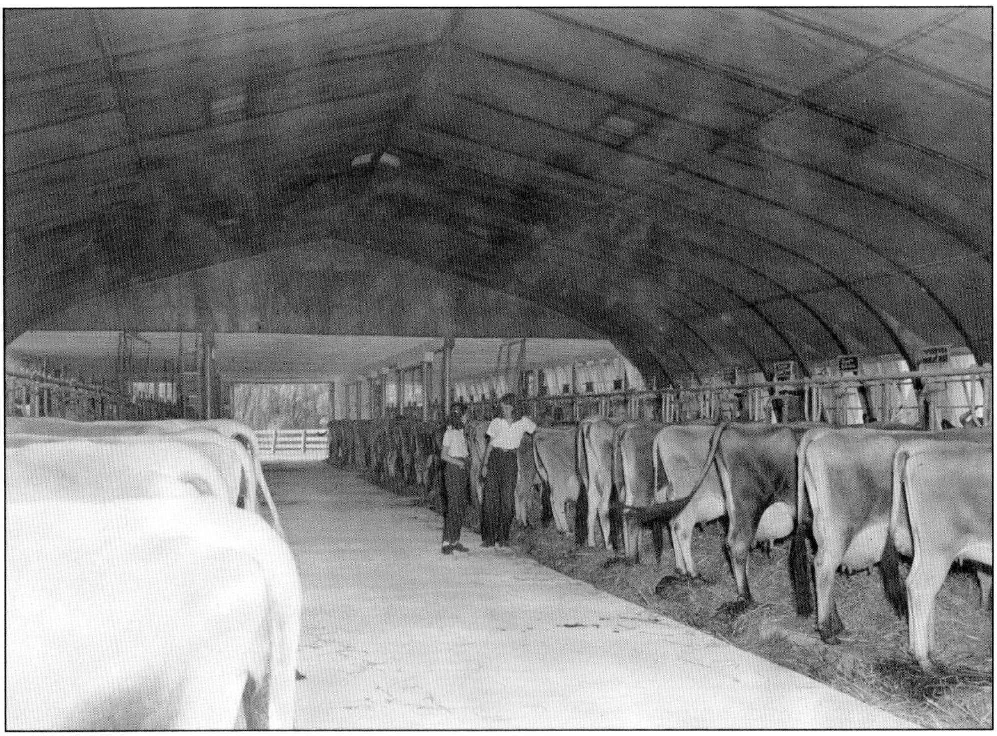

This inside view of the dairy barn at Holly Beach Farm, owned by the Labrot family in Whitehall near Annapolis, was taken around 1930. Two women are in the milk cow barn. The milk was sold to customers in the area. Of the five major breeds of dairy cattle, Jersey cattle produce milk that has the highest levels of protein and butterfat. (MSA.)

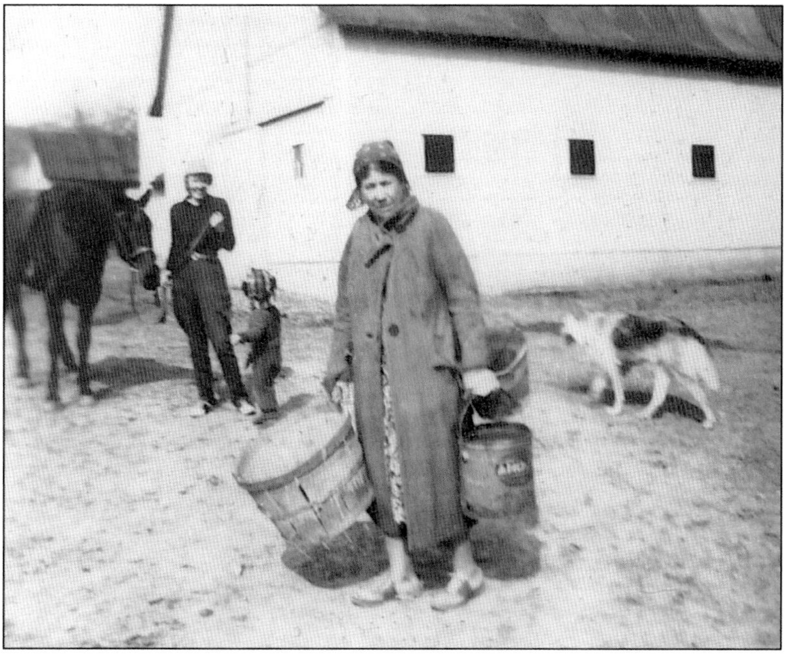

Charles B. Townshend had a 156-acre farm in Davidsonville. His wife, Rosa, is pictured outside their dairy barn in 1946. She had just fed the 50 to 60 hogs that were raised for market. In addition to hogs, they also had a dairy herd and raised horses, corn, and tobacco. In the background is her daughter Frances with one of her horses. (Scible.)

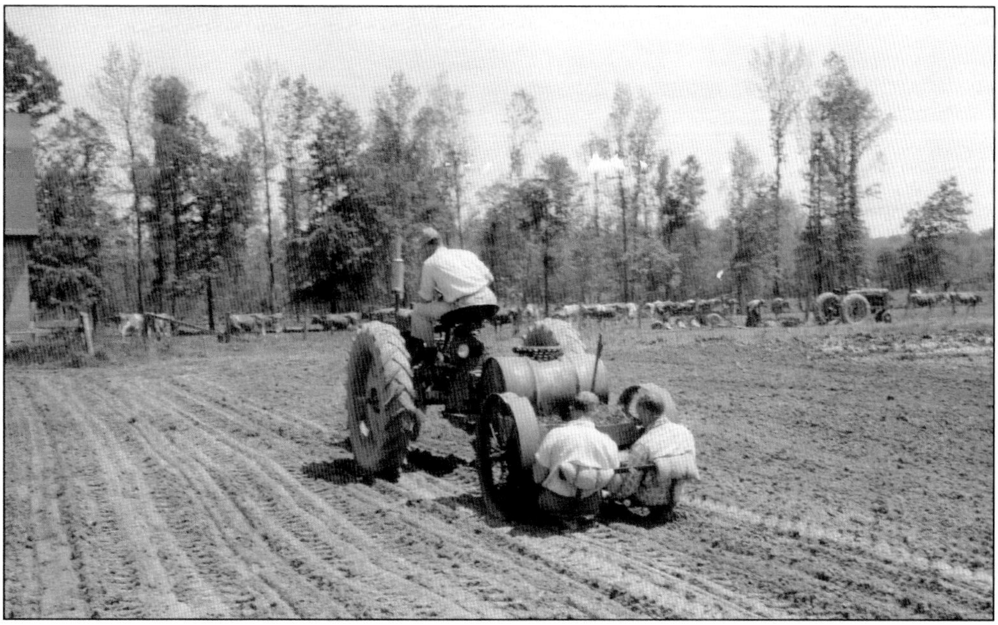

Folger "Mack" Ridout plants Midland Bermuda grass on his Woodlyn Dairy Farm in St. Margarets in 1958. The grass was used for pasture improvement and nursery stock on the farm. Orlando Ridout II began the dairy in 1920. Orlando Ridout III raised cows for milk and grew five acres of kale and turnips in the winter and sugar corn and tomatoes in the summer. The cattle were sold in 1962. (AASCD.)

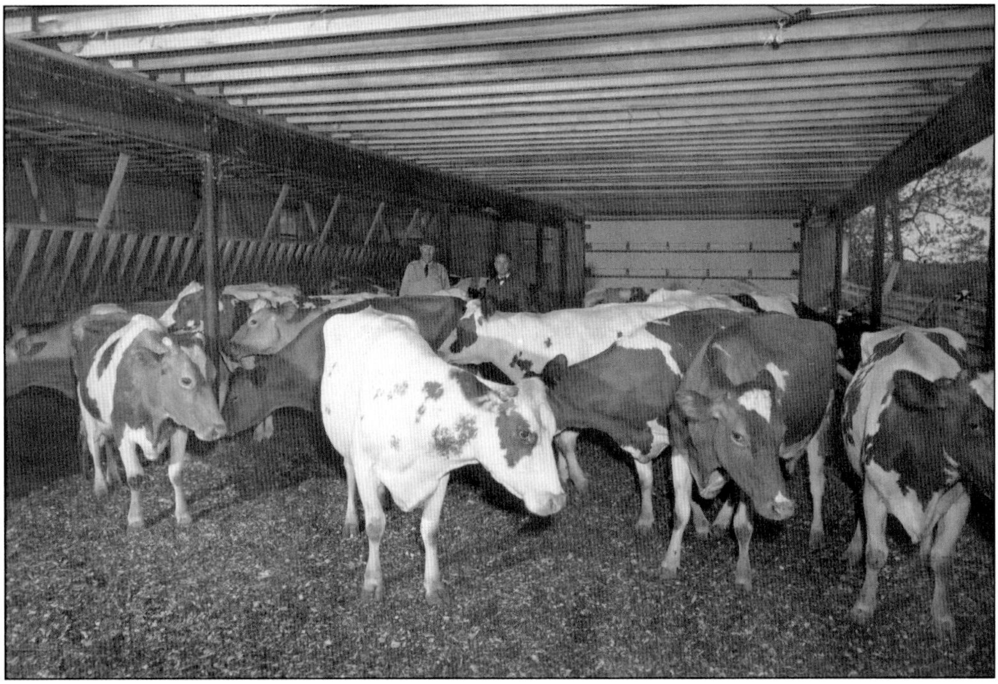

Francis E. Gardiner Sr. views his 120-head herd of Ayrshire dairy cattle in the feeding pen in the barn on his Wellaway Farm in Davidsonville in May 1952. About 75 milk cows were milked twice per day, and the milk was sold to the Annapolis Dairy. During World War II, he purchased the farm and raised dairy cattle and tobacco. (AASCD.)

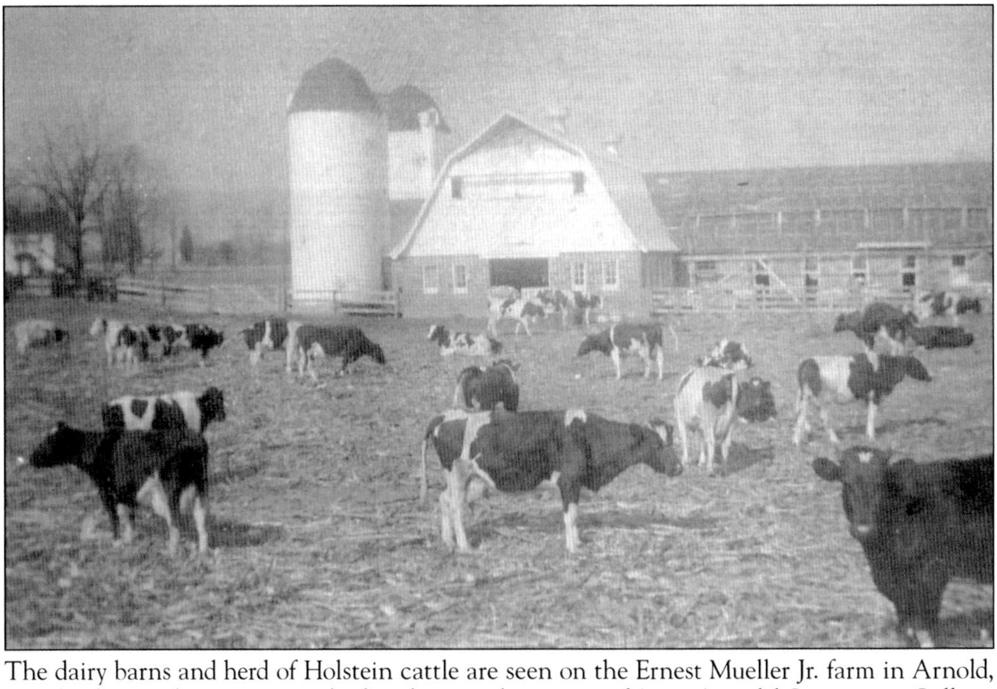

The dairy barns and herd of Holstein cattle are seen on the Ernest Mueller Jr. farm in Arnold, Maryland, around 1950. In 1961, the farm became the campus of Anne Arundel Community College. The barn is still used on the campus and can be seen from College Parkway. (Mueller.)

Edward S. Shepherd is seen with cattle on Oak Crest Farm on Solomons Island Road in Lothian in 1950. He was the son of Edgar Shepherd and raised tobacco, hay, cattle, pigs, and chickens. (Ernest Shepherd III.)

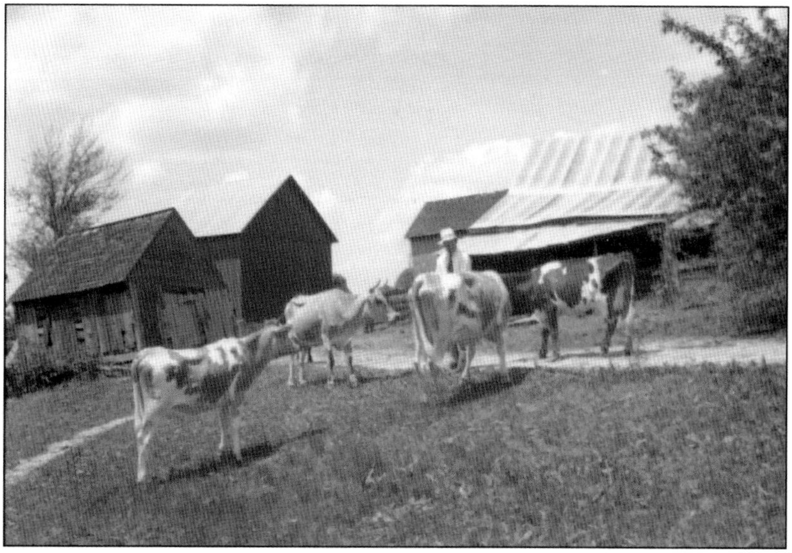

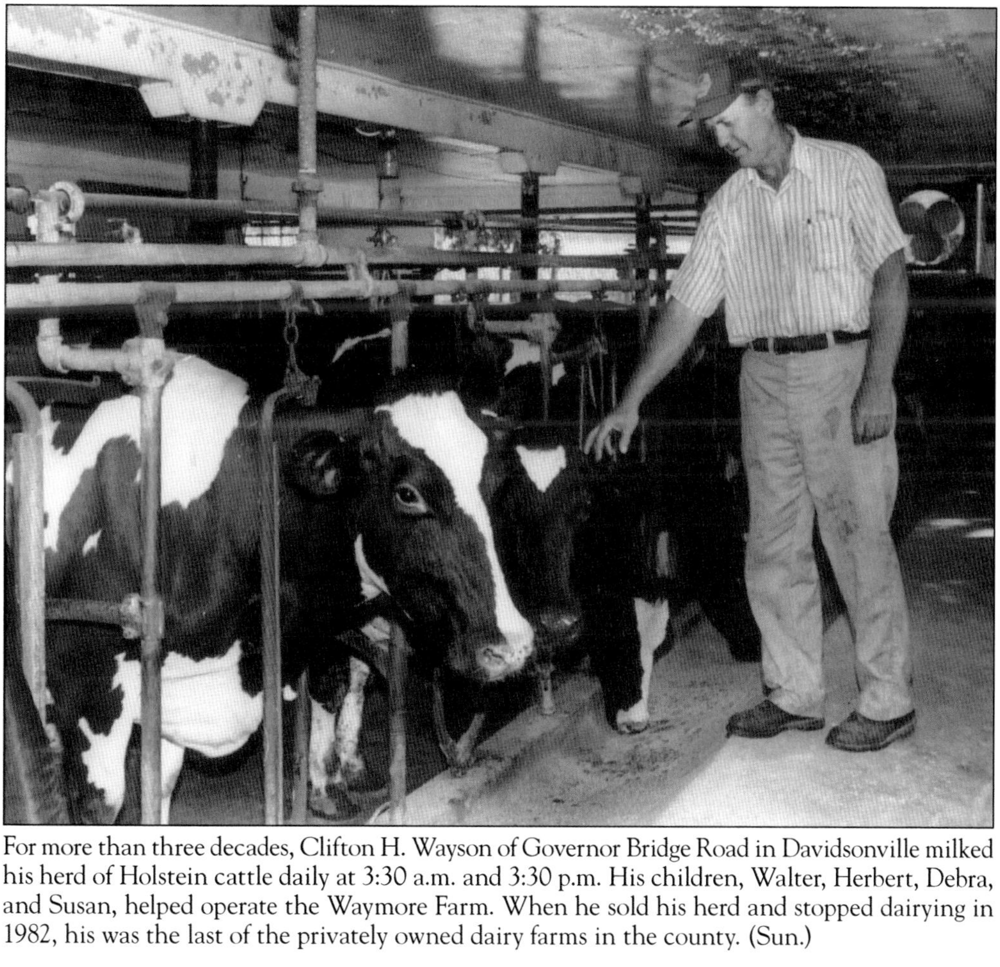

For more than three decades, Clifton H. Wayson of Governor Bridge Road in Davidsonville milked his herd of Holstein cattle daily at 3:30 a.m. and 3:30 p.m. His children, Walter, Herbert, Debra, and Susan, helped operate the Waymore Farm. When he sold his herd and stopped dairying in 1982, his was the last of the privately owned dairy farms in the county. (Sun.)

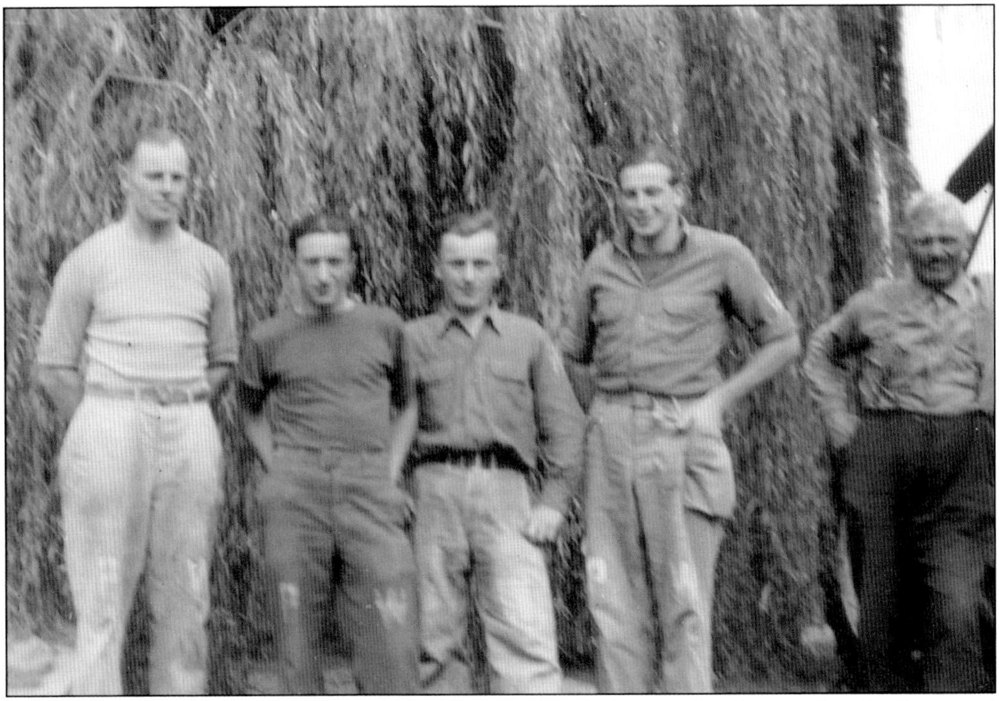

Joseph Bottner (right) is pictured with German prisoners of war on his farm in Gambrills around 1945. During World War II there was a shortage of farm laborers, so local farmers would drive to Fort Meade, where the prisoners were kept. They were picked up in the morning, assigned to work in the fields, fed, and returned in the evening. On the Bottner farm, they worked with tobacco and helped with hay. (Anderson.)

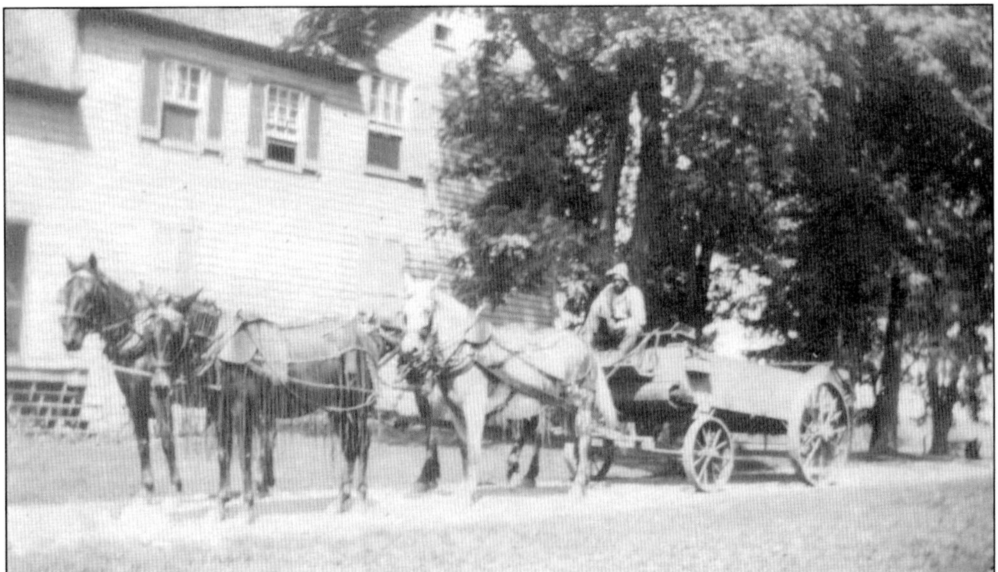

A team of horses pulls a manure spreader on the Addison Hodges farm in Davidsonville around 1910. Manure was gathered in the barns and then spread onto the fields to infuse organic matter and nutrients into the soil. (Mary Jay Davis.)

Four

Grain and Forage Crops

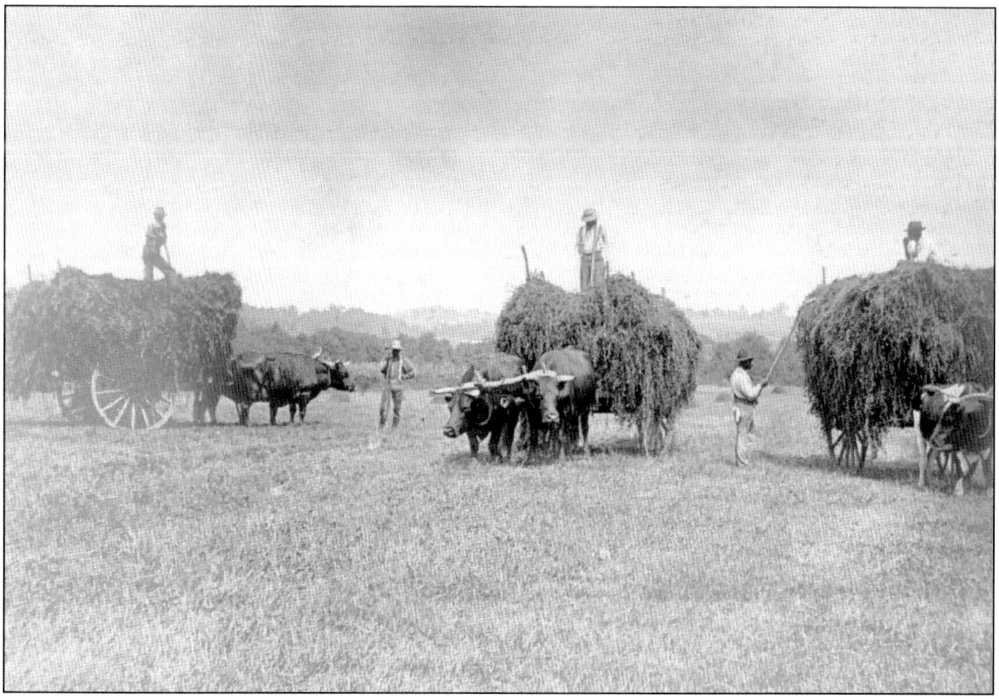

Ox-drawn carts haul hay from the front field on Ivy Neck Farm in West River around 1880. The hay is harvested and dried in the field before being loaded onto the wagons for storage in the barn. The hay provides feed for the livestock raised on the farm. (Colhoun.)

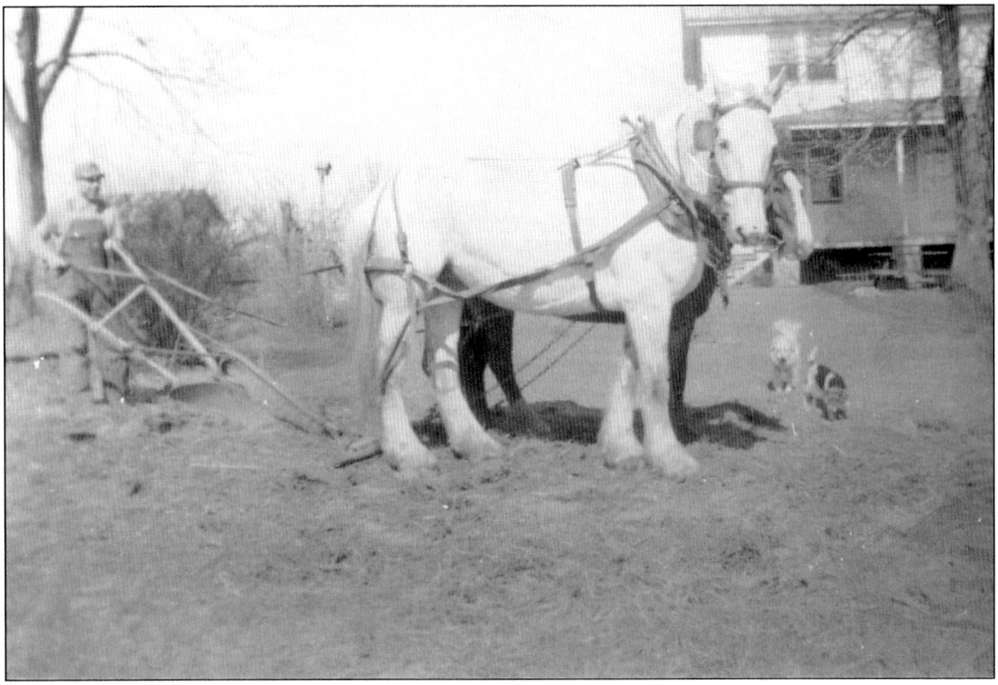

William P. Doepkens plows the home garden on the Henry Doepkens farm with his team of horses, Liz (white) and Bill, in 1942. Horses were used on the farm until the early 1950s. (Scible.)

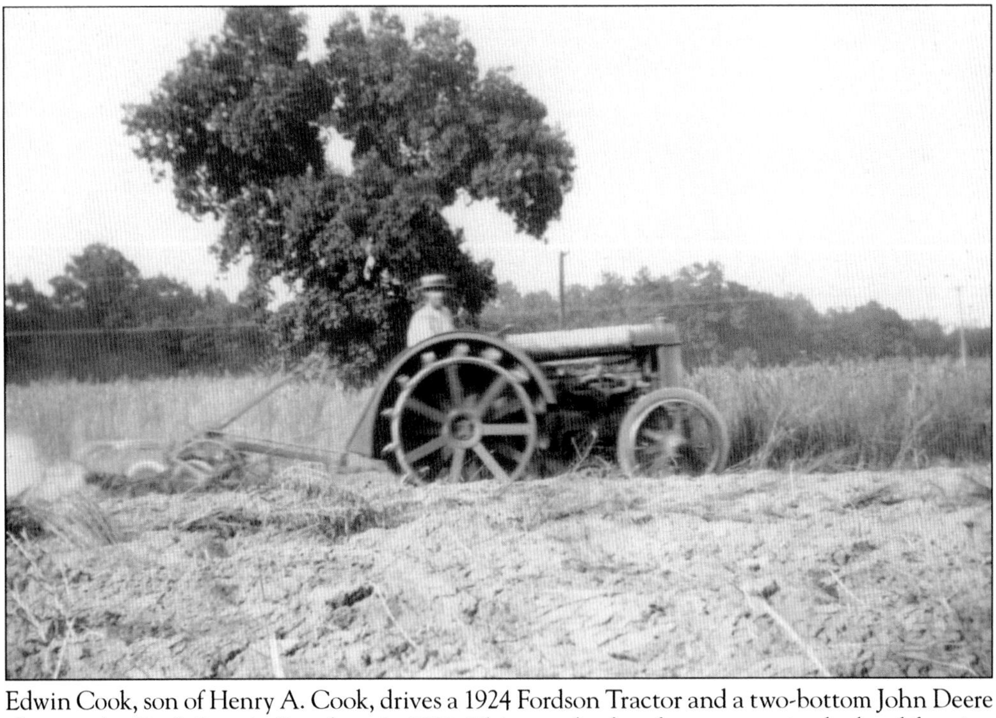

Edwin Cook, son of Henry A. Cook, drives a 1924 Fordson Tractor and a two-bottom John Deere plow on the Cook farm in Pasadena in 1932. This was the first farm tractor in the local farming community. Purchased by Henry A. Cook, it was used for many years on the Cook farm, primarily for land preparation, plowing, and discing. (Calvert.)

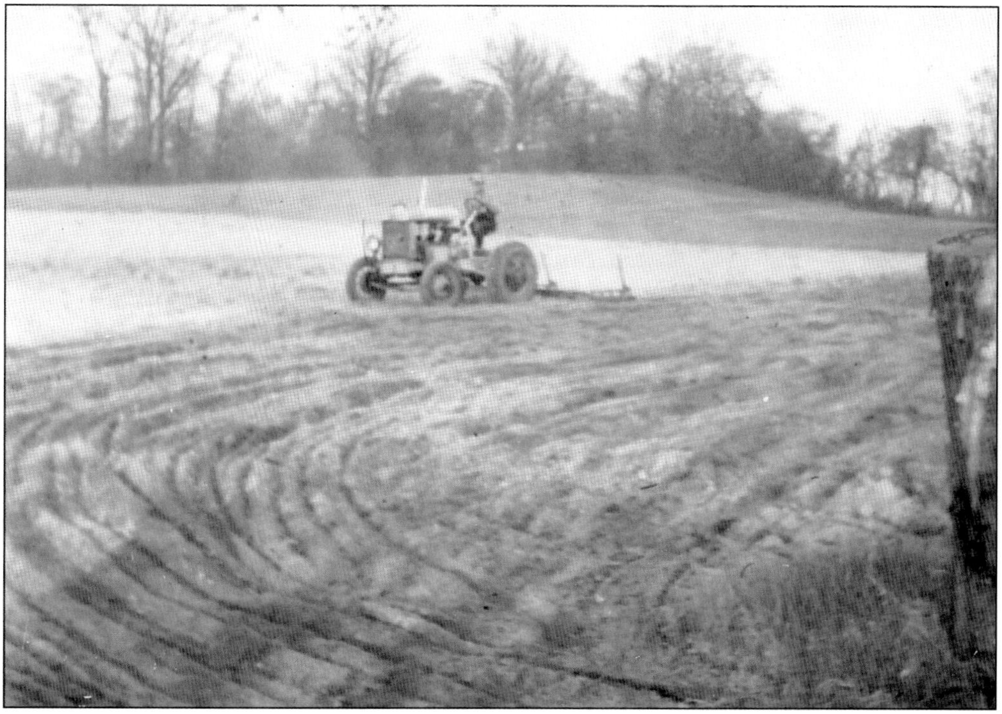

Edwin Calvert harrows a field with a homemade tractor in 1947. During World War II, tractors were not available for purchase, so farmers were creative and made some of the equipment they needed. This tractor was made using a Ford Model A truck, onto which a radiator and heavy-duty rear axle were added to provide power. (Calvert.)

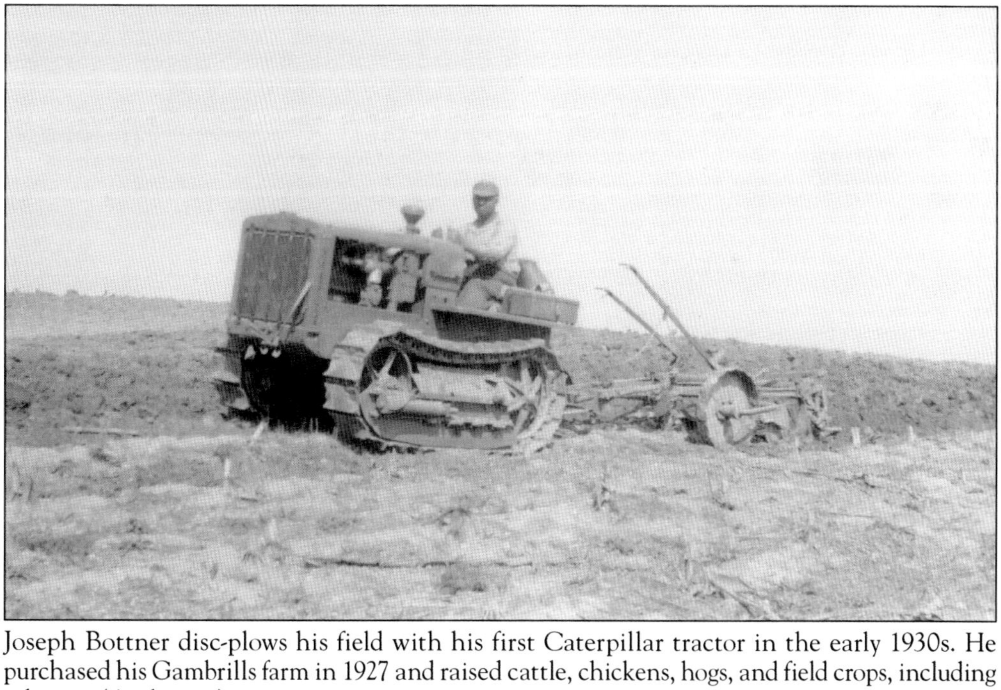

Joseph Bottner disc-plows his field with his first Caterpillar tractor in the early 1930s. He purchased his Gambrills farm in 1927 and raised cattle, chickens, hogs, and field crops, including tobacco. (Anderson.)

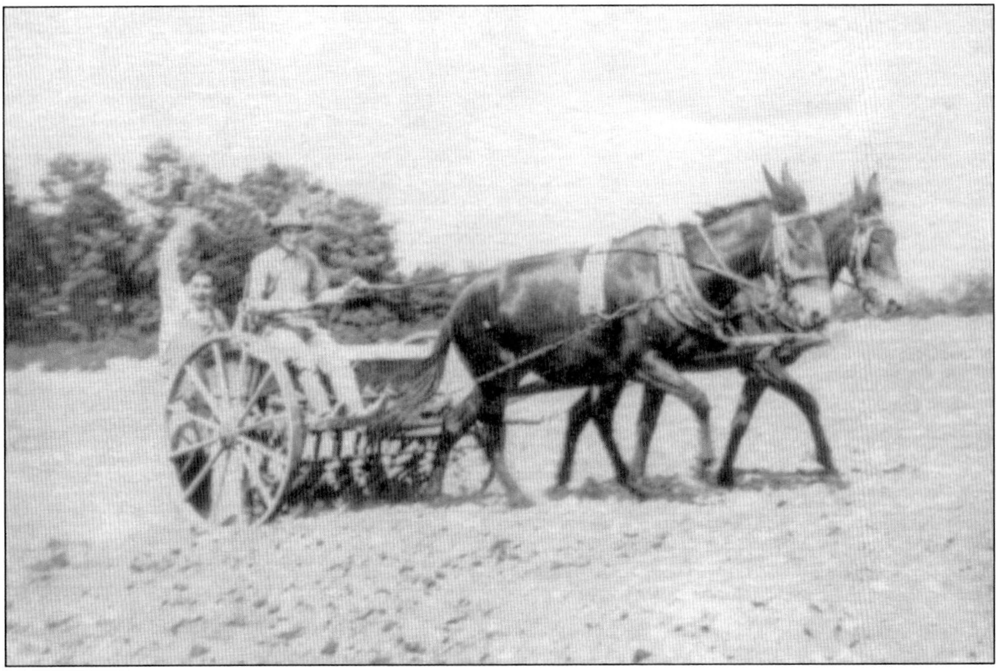

John Gertz leads the mules while Edward Gertz rides on the back of the grain drill as they plant wheat for a cover crop on the Gertz farm in 1924. (Gertz.)

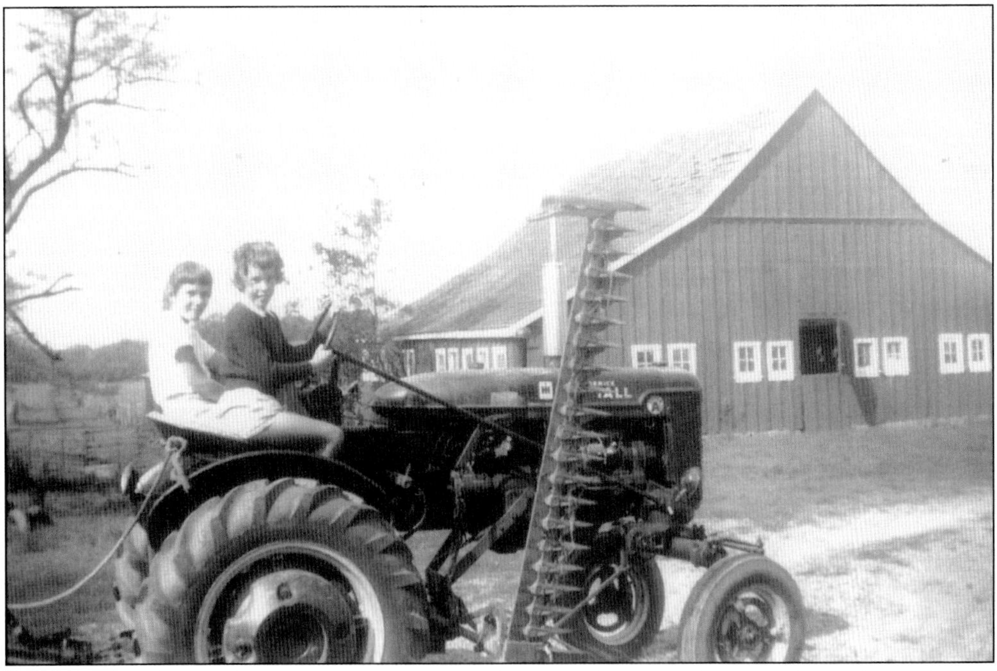

This barn on the Goshen Farm in Cape St. Claire was home to dairy cattle in the early part of the 20th century and then to beef cattle when purchased by Dr. Morris L. Radoff in 1942. In 1970, his nieces, Olivia (left) and Linda Radoff (right), sit on their uncle's Farmall tractor, which has a sickle mower attached. (Photograph by May Radoff, courtesy of the Goshen Farm Preservation Society.)

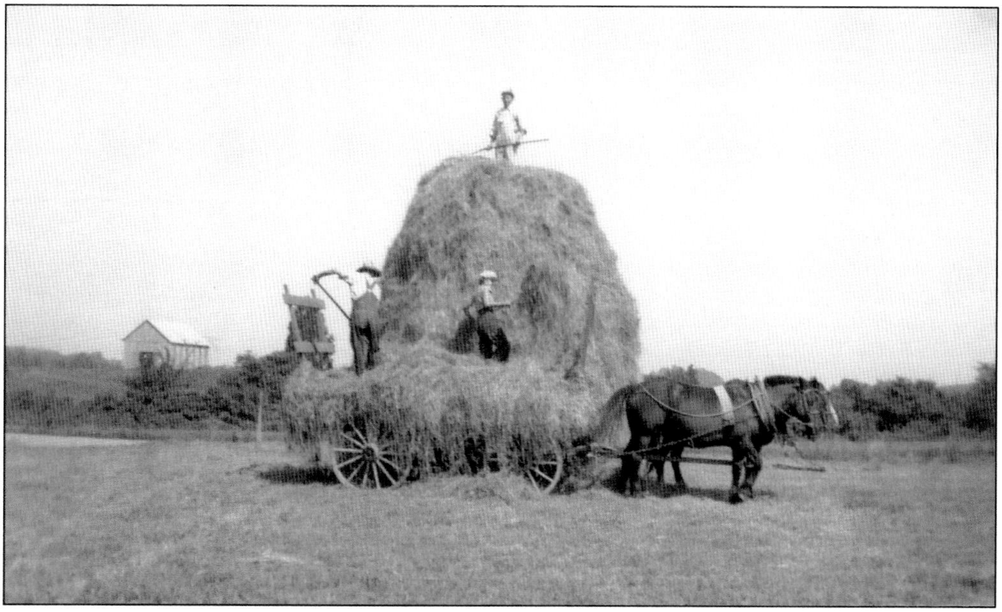

William P. Doepkens (left) and Teresa Doepkens (right) stand on the hay wagon as they fork hay onto a haystack in 1938. Their dad, Henry Doepkens, is on top of the stack of hay. He was an emigrant from Twistringen, Germany, and bought his first farm in the Davidsonville area in 1906. In 1922, he bought his 210-acre Gambrills farm. (Scible.)

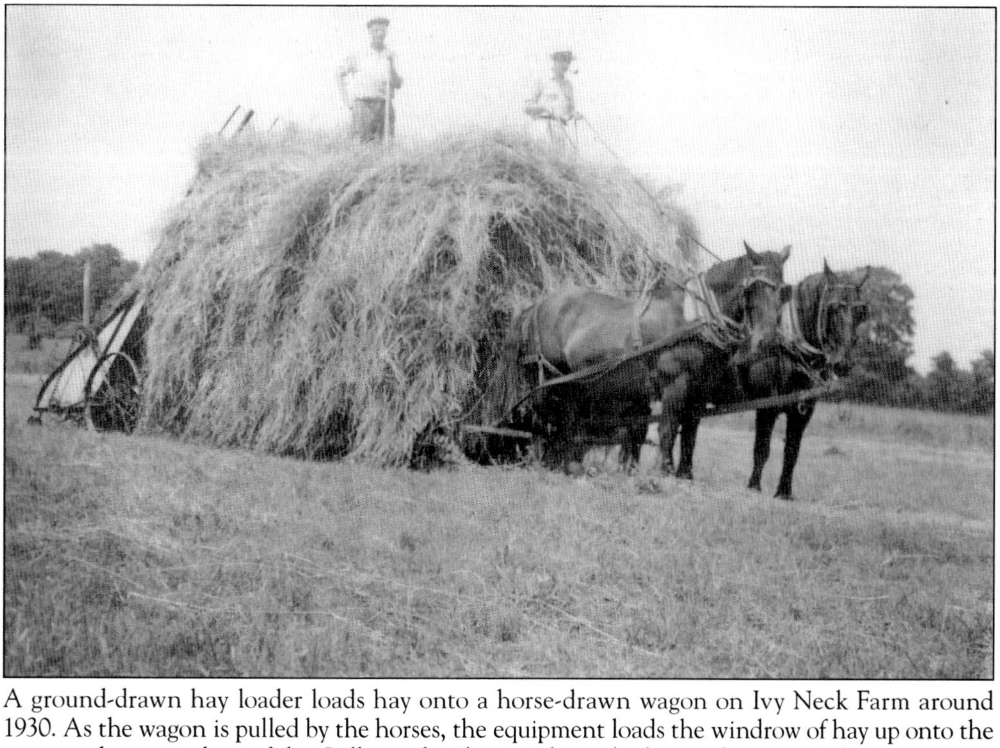

A ground-drawn hay loader loads hay onto a horse-drawn wagon on Ivy Neck Farm around 1930. As the wagon is pulled by the horses, the equipment loads the windrow of hay up onto the wagon, where members of the Colhoun family spread out the hay and prepare it for transport to the barn. (Colhoun.)

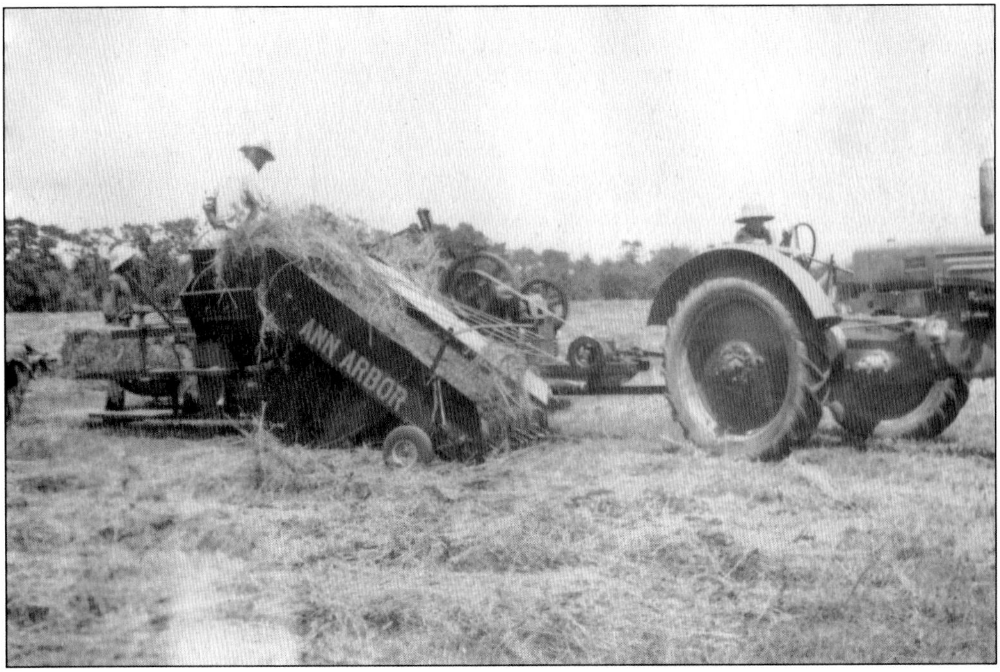

The Colhoun family bales hay at Ivy Neck Farm with an Ann Arbor baler in 1940. This type of baler was built by the Ann Arbor Machine Company in Illinois, which manufactured the first tractor-pulled pickup baler in 1932. Two men were needed on the back of the machine to tie wire around the compressed 16-by-18-inch bale. (Colhoun.)

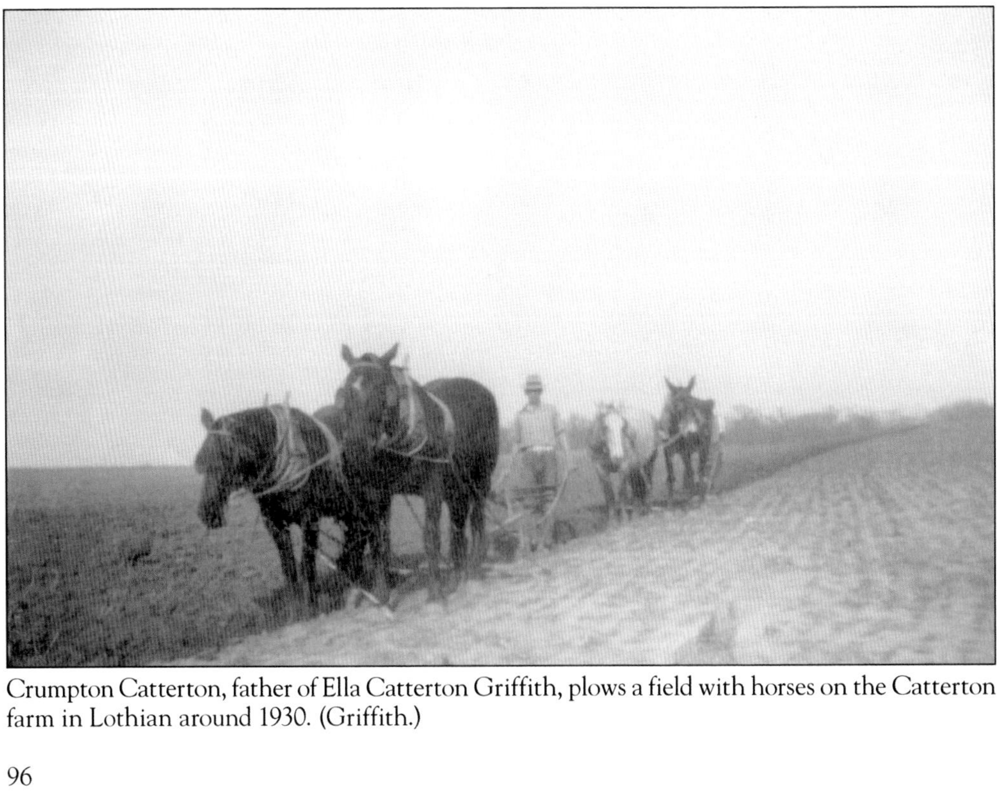

Crumpton Catterton, father of Ella Catterton Griffith, plows a field with horses on the Catterton farm in Lothian around 1930. (Griffith.)

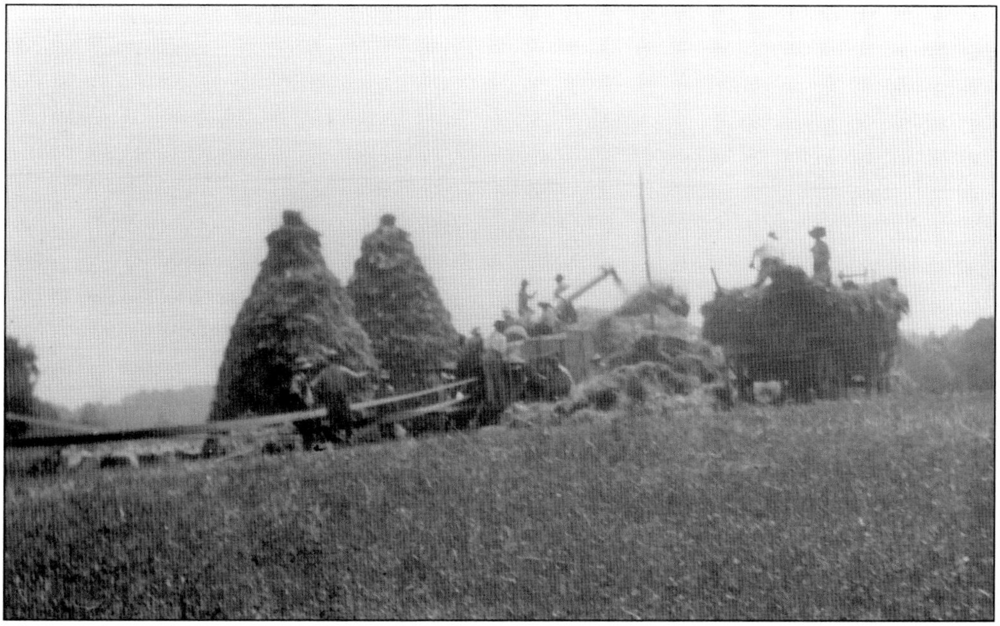

The Bunker Hill Farm, located in Millersville, was owned by Judge William Baldwin. Wheat is threshed on the farm around 1920. Farm workers on the wagon loaded with bundles of wheat load the thresher. Two tall piles of wheat straw are on the left. The farm was operated by Summerfield Baldwin. The Baldwin family owned Warren Mill and Savage Mill, which milled flour in Baltimore County. (MSA.)

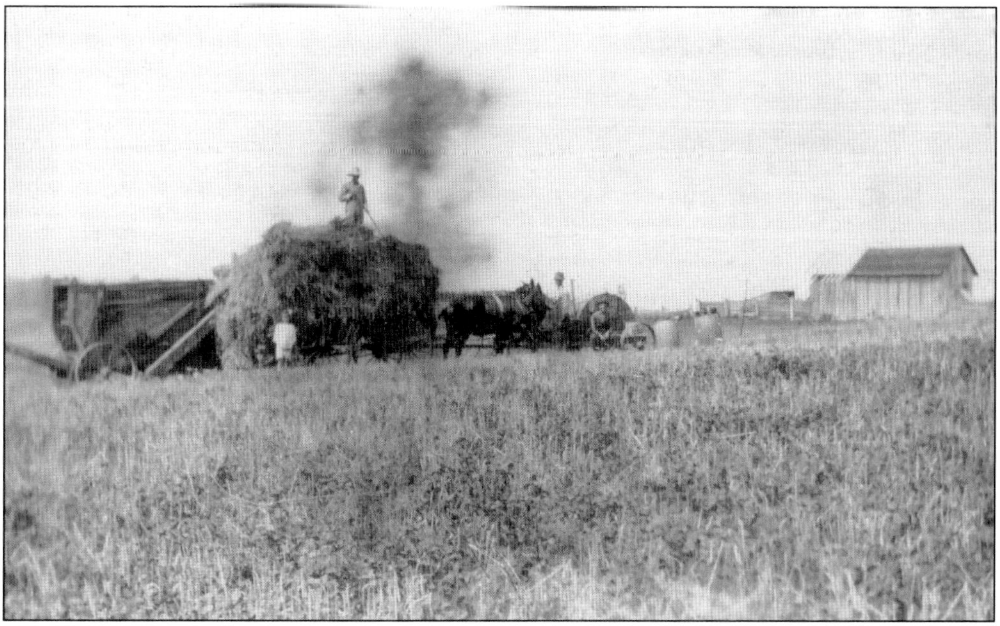

Farmers thresh wheat on the Gerstner farm in Gambrills around 1920. John Gerstner came to the United States from Obertshausen, Germany, in 1886 and was a baker in Baltimore for 25 years before he purchased his farm in 1910. Otto J. Gerstner is on the loaded hay wagon on the 356-acre Gerstner farm in Gambrills. They raised Guernsey and Holstein cattle and sold the cream, milk, and butter in Baltimore. (Anderson.)

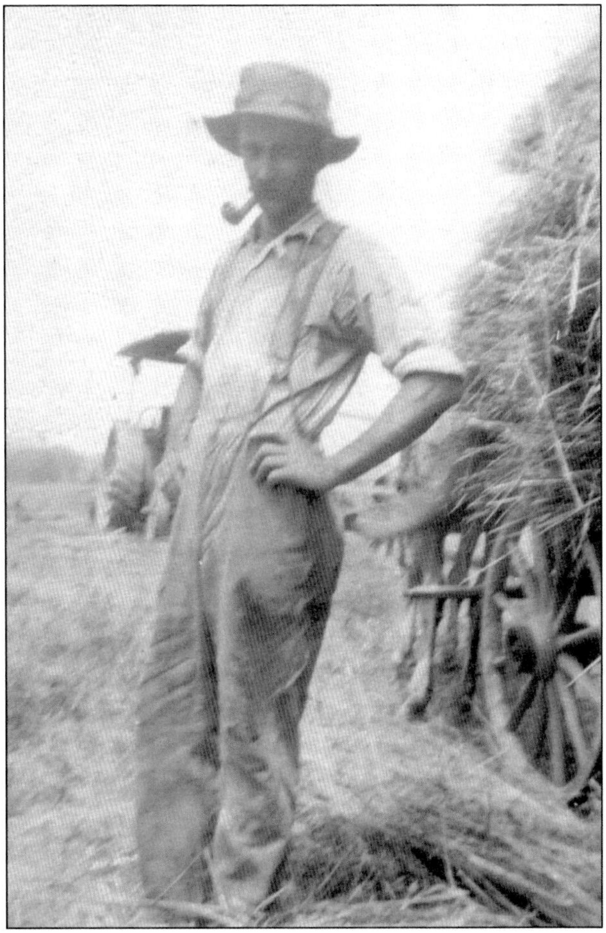

William O. Tucker Sr. was born around 1880 in the county and was a prosperous farmer and miller. He would take his threshing machine around to farms in the Davidsonville area to thresh their grain crops. His son, William O. Tucker Jr., did custom combining for farmers in the 1970s and 1980s. This photograph is from around 1910. (Chase.)

Wheat is threshed on the Henry Doepkens Farm around 1930. Wagons of wheat bundles are brought from the field to where the threshing machine is set up. The wheat is forked into the threshing machine, where the grain is separated from the straw. Bags of wheat sit on the ground to the left of the thresher, and a large pile of straw is in the background. (Scible.)

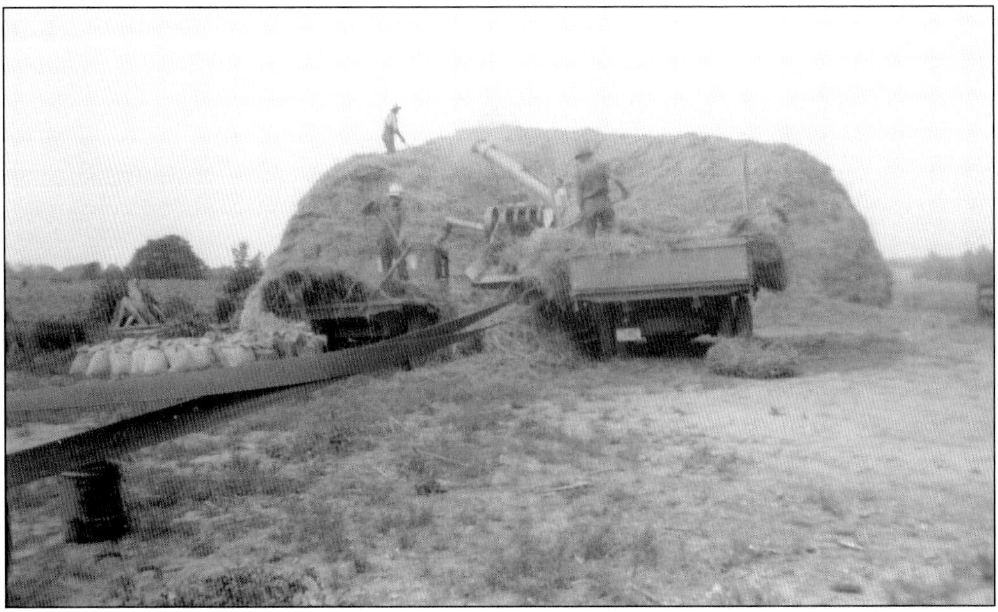

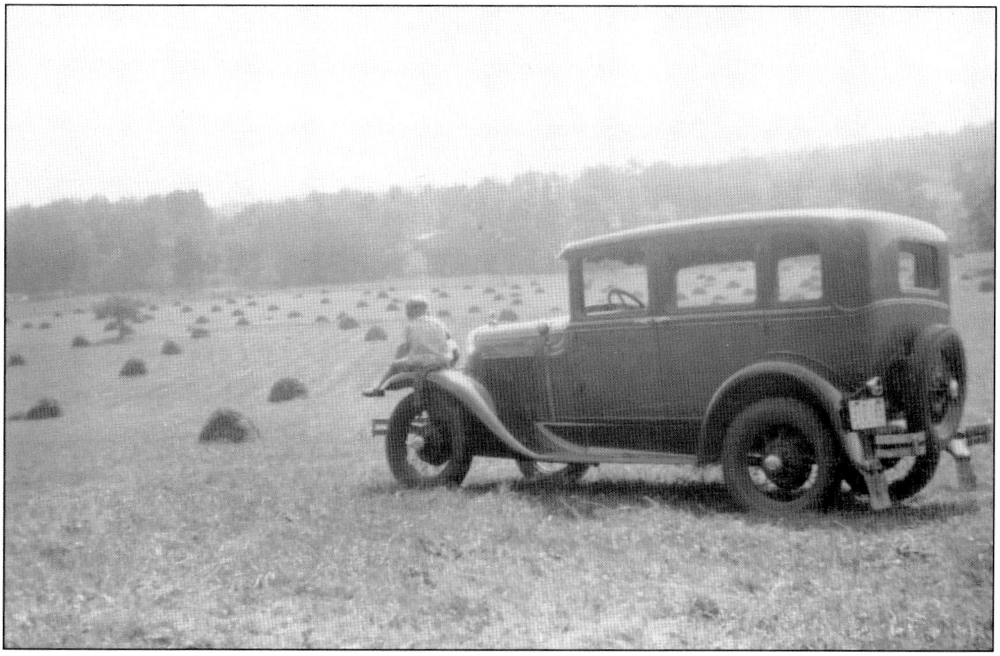

Shocks of hay dry on the Calvert farm in Pasadena in 1935. Produce fields harvested during the summer months were planted with a crop of crab grass that was mowed for hay in the late fall. The fallow fields were planted with cowpeas that were cut for hay. The cowpeas added nitrogen to fields where they were grown. (Calvert.)

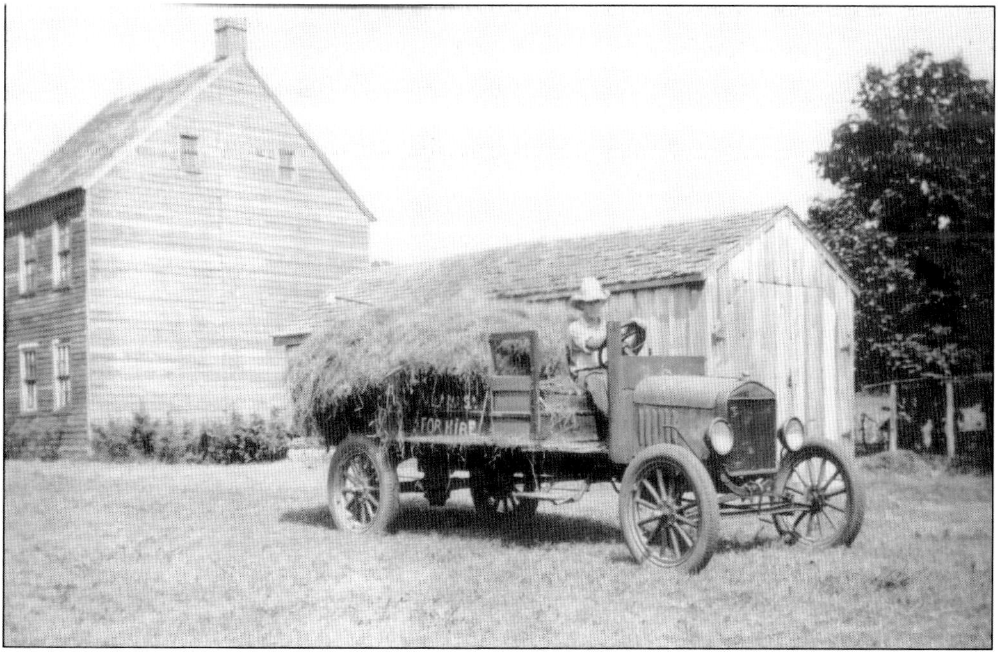

James B. Calvert drives a 1923 Ford Model T truck loaded with hay on the August Calvert farm in 1932. Purchased in 1931 and used for several years, this was the first truck on the Calvert farm. The hay was hauled to the barn or stacked in haystacks and saved for winter livestock feed. (Calvert.)

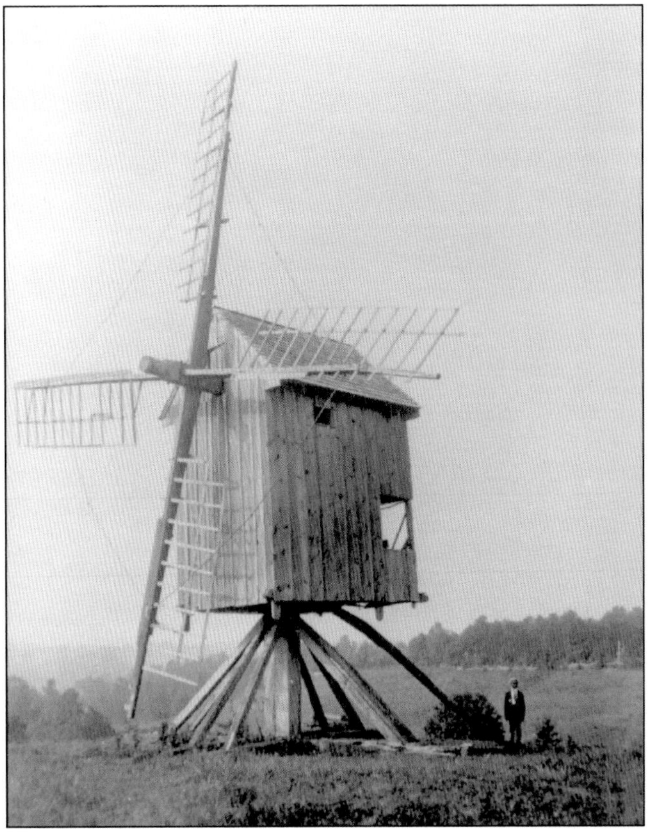

In 1905, this windmill operated by Thomas Mullen, an African American, was the last operating windmill in the county. When the wind was favorable, it was used to grind grain for man and farm animals. The mill was located on the Holly Hill Farm in Friendship, which was part of a quarter plantation owned by Richard Harrison from 1679 until 1733. (MSA.)

John Ripple, farm owner and enrollee in the Veterans' Farm Training Program, examines his fields, showing contour strip cropping on his farm in Millersville in May 1952. The contour strips alternated with a cover crop of vetch and wheat and cultivated land ready to plant tobacco. (AASCD.)

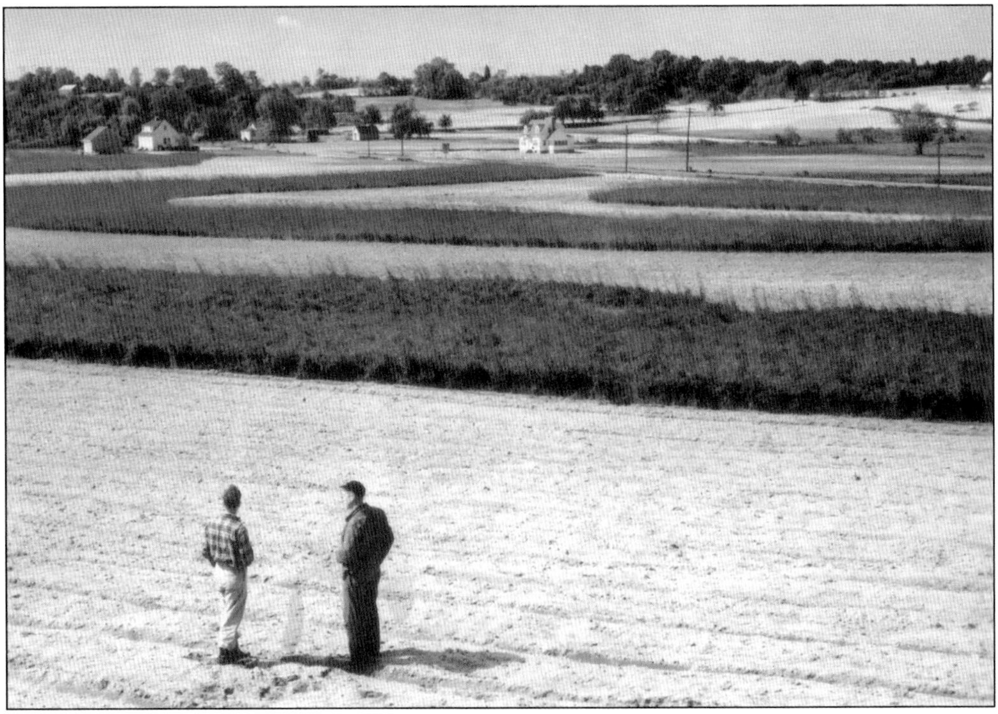

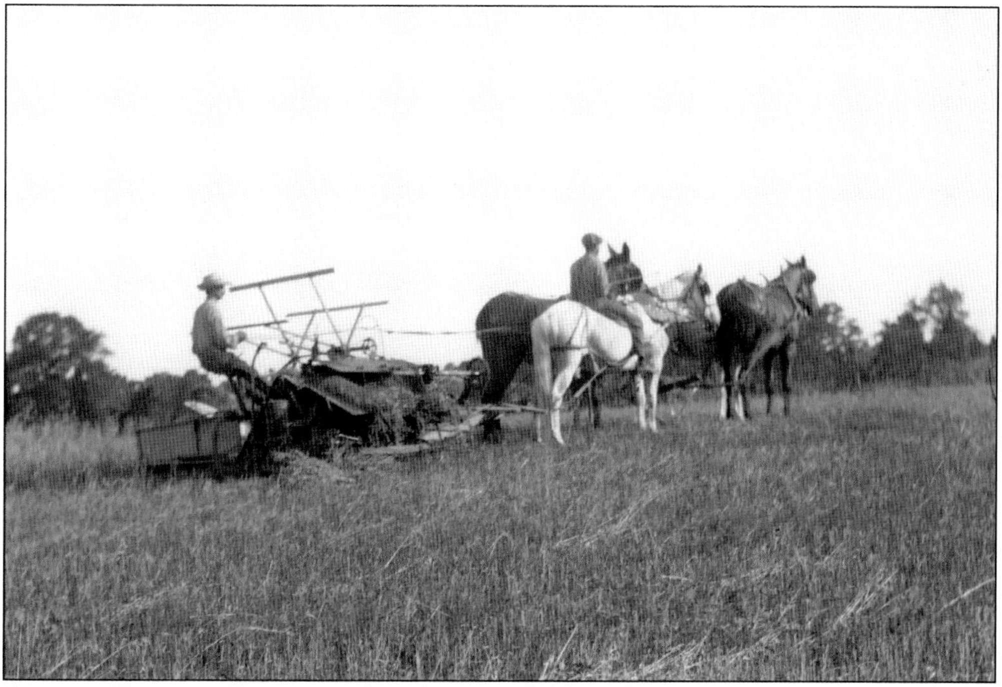

A team of horses pulls a reaper though an oat field on the Bunker Hill Farm, operated by Summerfield Baldwin, in Millersville around 1920. The reaper cut and bound the oats into shocks for later threshing on the farm. (MSA.)

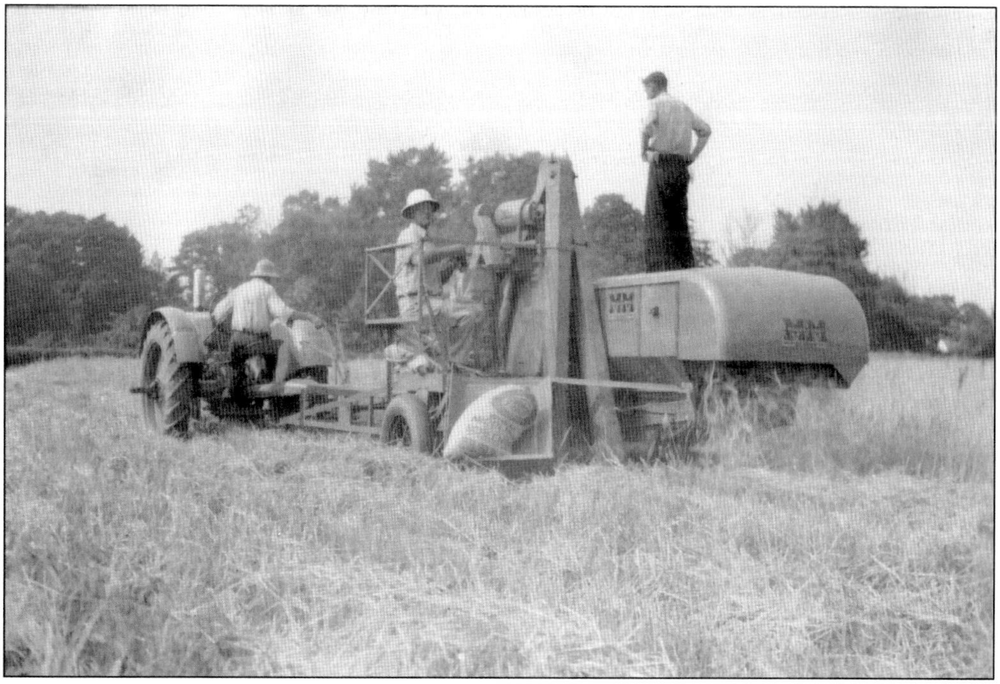

Wheat is combined on the Ivy Neck Farm using a Minneapolis Moline "MM" combine pulled by a tractor in 1940. The machine combines into a single operation three tasks that used to be done by separate machines: reaping, threshing, and winnowing. (Colhoun.)

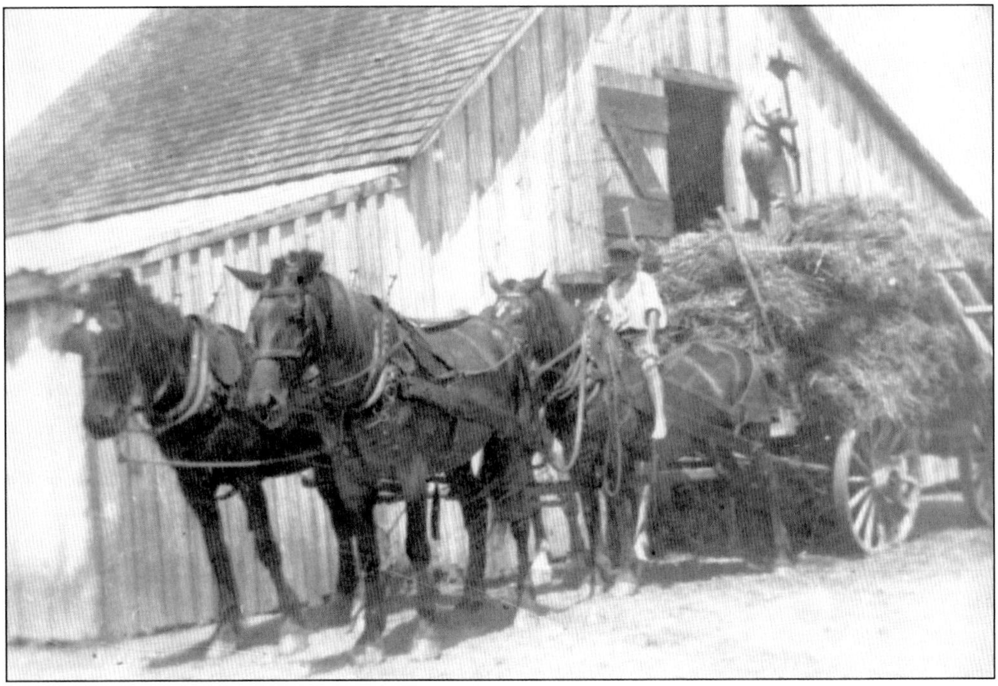

William O. Tucker Sr. unloads hay into the barn on his family farm as his son, William O. Tucker Jr., holds the horse reins around 1920. Pigs, turkeys, and ducks were raised for family consumption, but the Tuckers also raised barley, corn, hay, and show horses. By the late 1940s, the 100-acre Davidsonville farm had 35 dairy cows, which produced about 90 gallons of milk per day for the Annapolis Dairy. (Chase.)

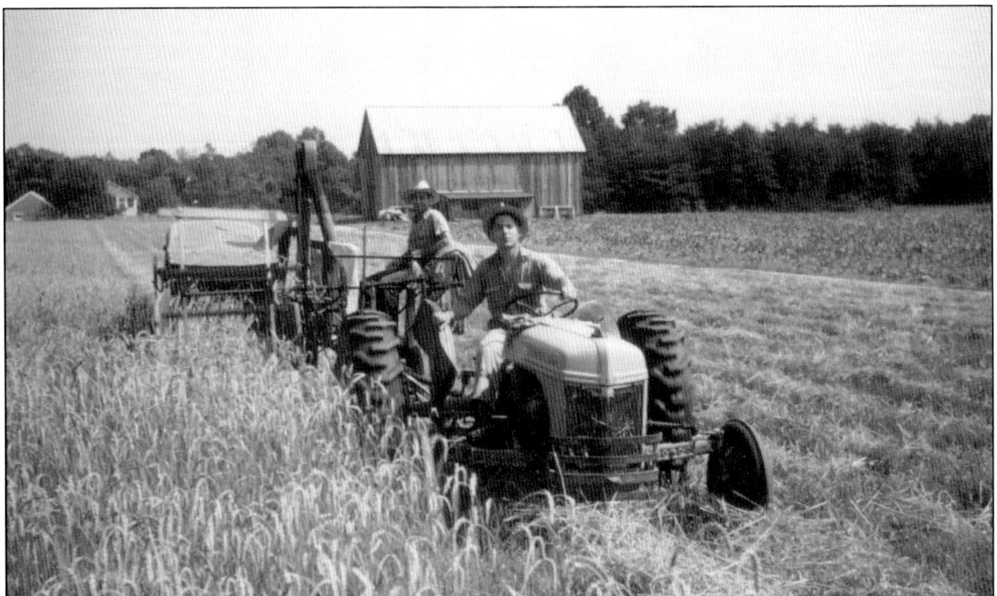

Jack Gertz drives the tractor while Edward Gertz rides on the combine bagging the cleaned grain around 1950. Edward Gertz purchased his farm on Patuxent River Road in Davidsonville in 1946. The tobacco barn in the background was built by George Bottner. (Lorraine Clow.)

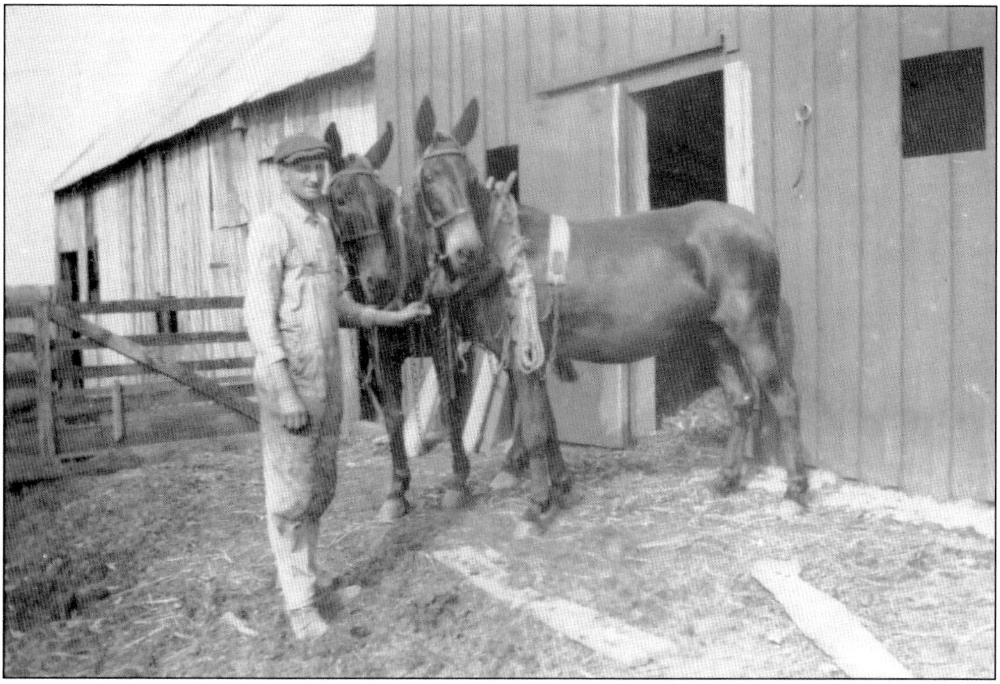

The Gerstner farm had about six mules that were used to work the fields. Shown is Otto J. Gerstner in 1929 with two of their mules outside their stable on the farm in Gambrills. (Anderson.)

A Cook farm laborer in Pasadena cuts off corn to make corn shocks in 1930. On the Calvert farm, the corn stalk was cut above the ear. After it was dried in stacks, the fodder was removed for animal feed. The ear corn cured on the stalks in the field, and in October, it was pulled, hauled to the barnyard, and eventually husked and used for feed. (Calvert.)

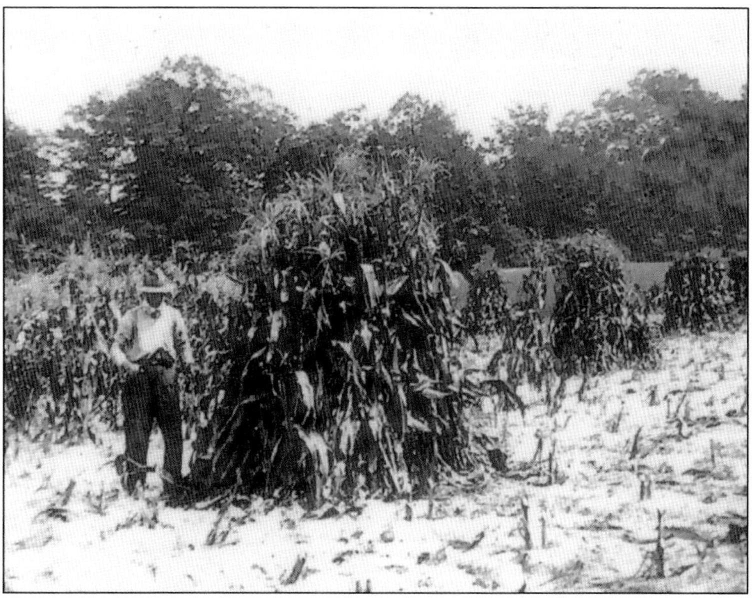

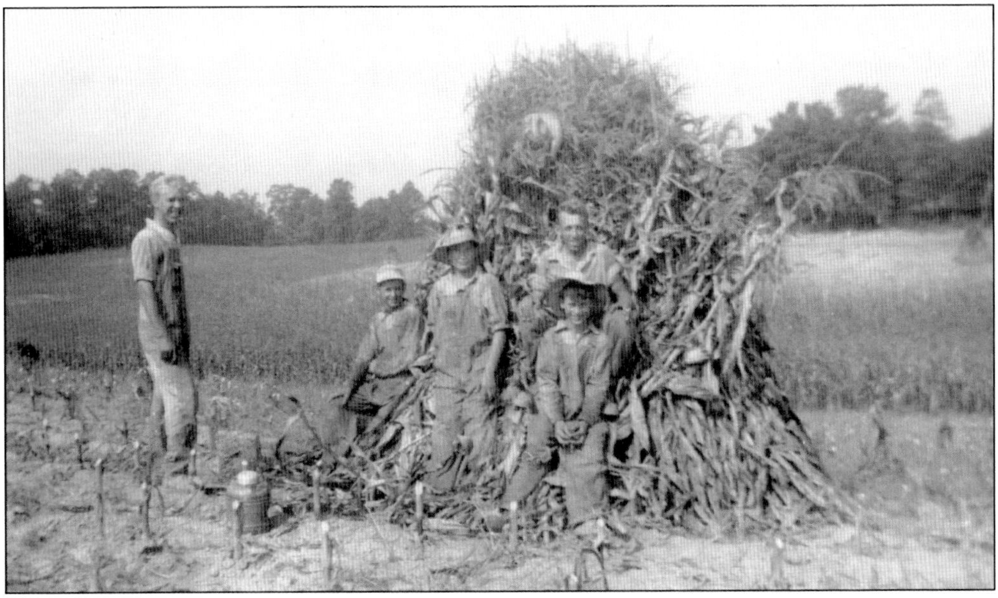

A harvest scene on the Henry Doepkens farm around 1940 shows, from left to right, William P. Doepkens, Michael Wagner, Nicholas Wagner, Charles Wagner, and Mathias Doepkens. According to Doepkens's book on Middle Plantation, 1947 was the last year corn was put up in shocks on the farm. (Scible.)

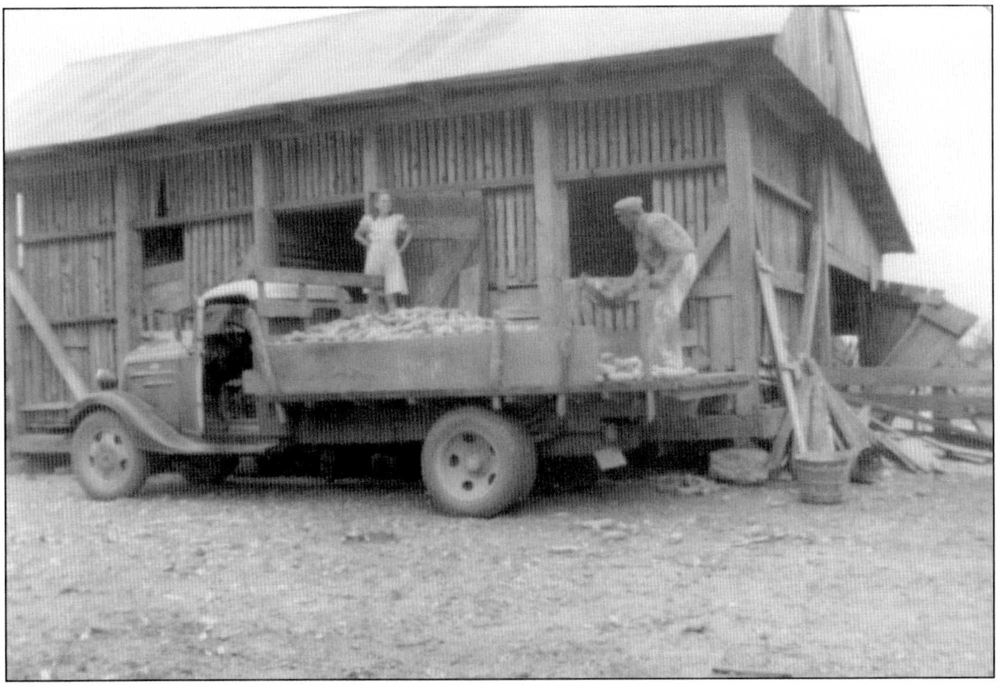

William P. Doepkens shovels ear corn from the family's farm truck into their corn house while his sister, Mary Helen Doepkens, looks on around 1940. This corn house was replaced with a newer one around 1949. The corn would be used on the Henry Doepkens farm for animal feed. (Scible.)

Farm silos are used to store corn or hay silage, which is fed to ruminant livestock animals during the winter. In August 1975, Frederick Doepkens uses a rake to unload the silage into a trough that will auger the chopped-up corn to a blower that will fill the silo. His dad, William P. Doepkens, is on the top of the forage wagon. (Salyers.)

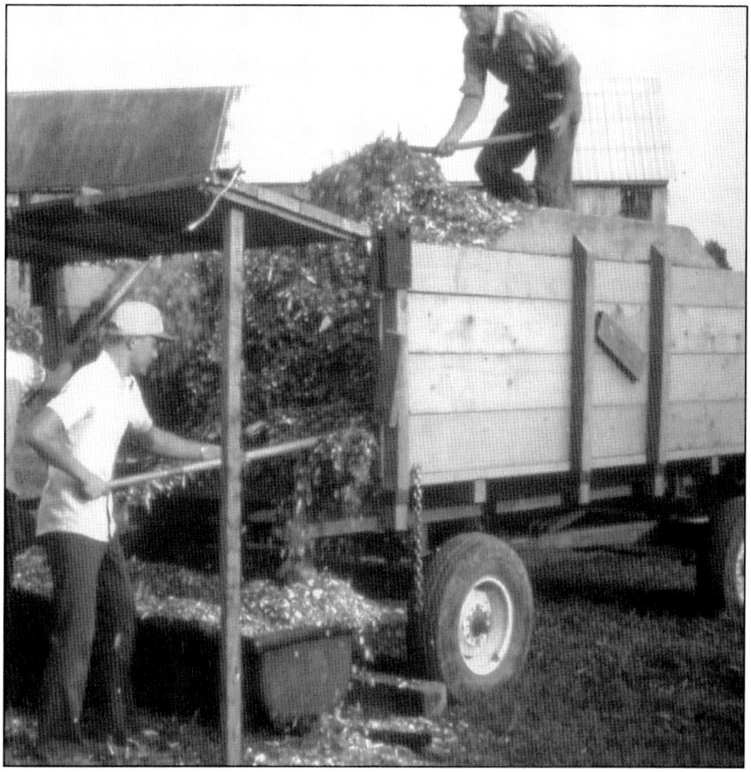

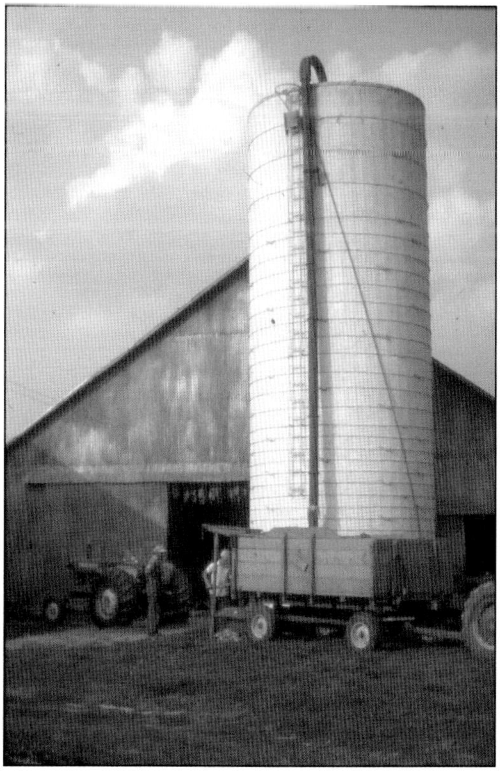

A long pipe is attached to the outside of the silo, and a blower is used to blow the silage up the 10-inch-diameter pipe to the top of the silo in August 1975. Forage harvesting equipment from Benjamin Watkins III worked for several days to chop the corn and fill the silo. Once fermented, the silage was fed to the herd of beef cattle on the Doepkens farm. (Salyers.)

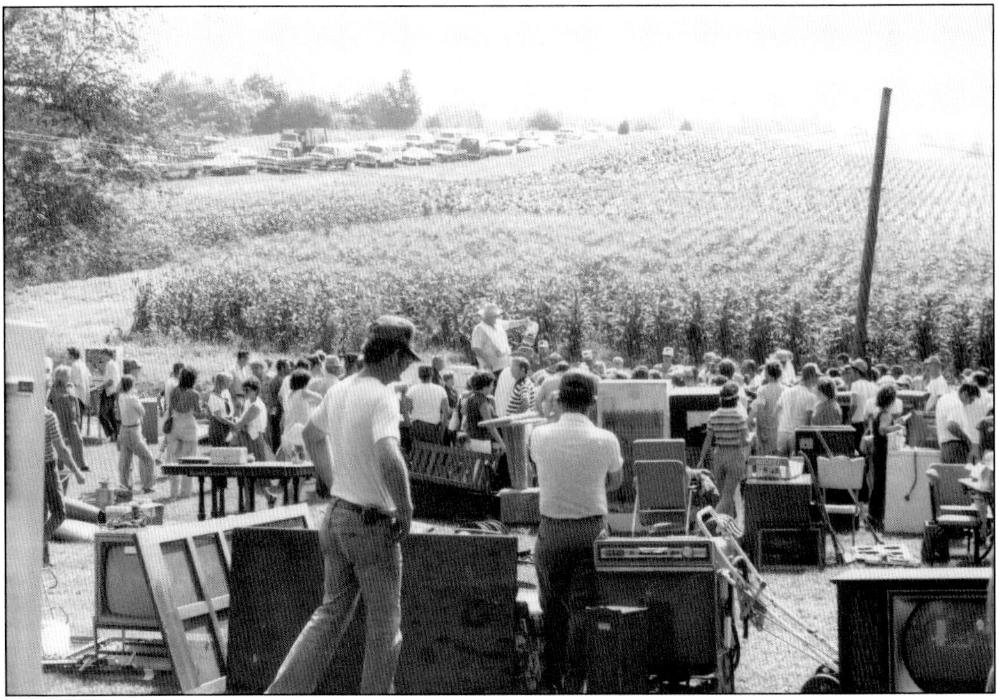

For over 30 years, the Davidsonville Ruritan held an annual auction on the Locust Farm, owned by Benjamin Watkins III, in Davidsonville. Robert H. Campbell auctions farm machinery, antiques, and a wide variety of household items to help benefit community programs sponsored by the Ruritan club in 1990. From 1962 until 1997, the auctions were an annual tradition for the community. (Davidsonville Ruritan Club.)

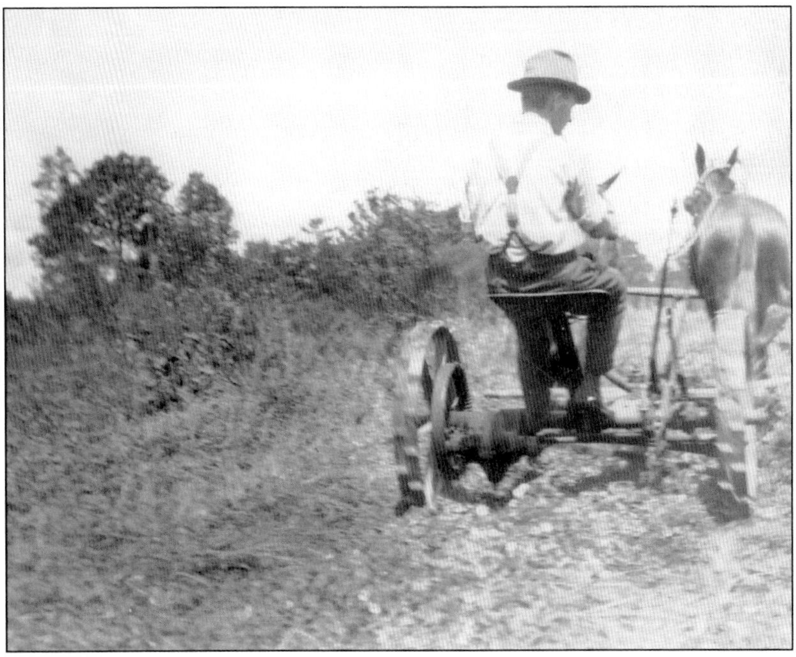

John Gertz mows hay with mules and a sickle mower on his family farm in Waterbury–Crownsville around 1910. As a child, John emigrated from Wloclawek, Russian Poland, with his family in 1902. (Gertz.)

Five

Horticulture and Nursery Crops

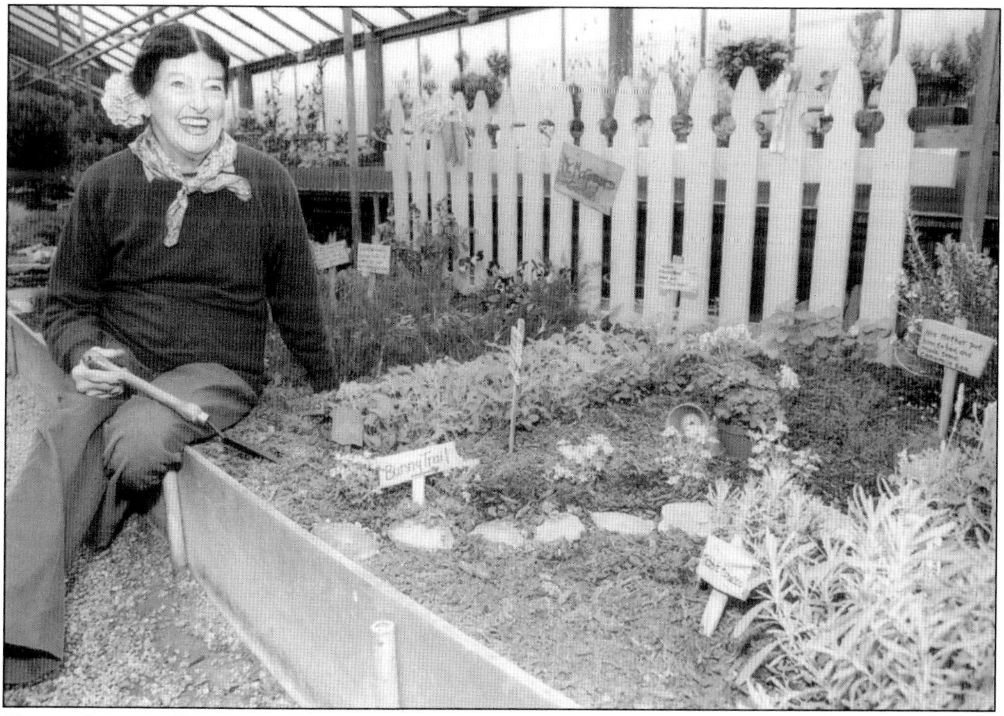

Horticulture and nursery crops have become big business in Anne Arundel County. In 1961, Hildreth K. Morton started Bittersweet Hill Nurseries on Governor Bridge Road in Davidsonville. She started raising tomato plants in a solitary greenhouse and quickly expanded her operation to herbs, water lilies, and many other flower and vegetable bedding plants. In this 1997 photograph, she cares for her herbs. (Capital.)

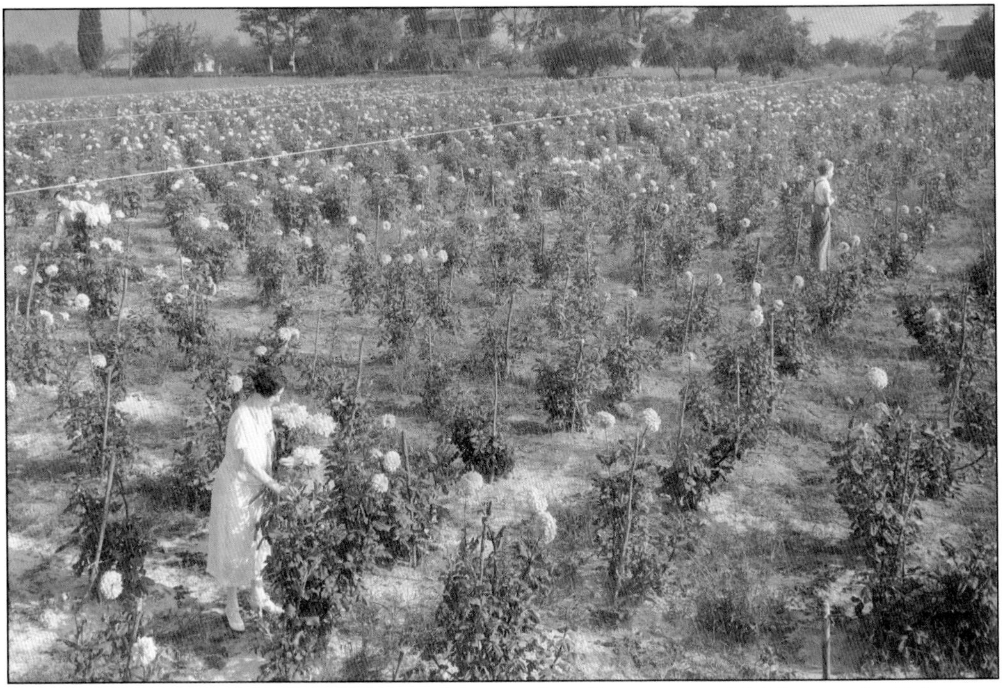

Peter L. Schramm and Frances Schramm harvest their flower crop of dahlias on the Schramm farm in Pasadena in late September 1931. They were one of the largest flower growers on the east coast and sold most of their flowers to markets in Baltimore. (Sun.)

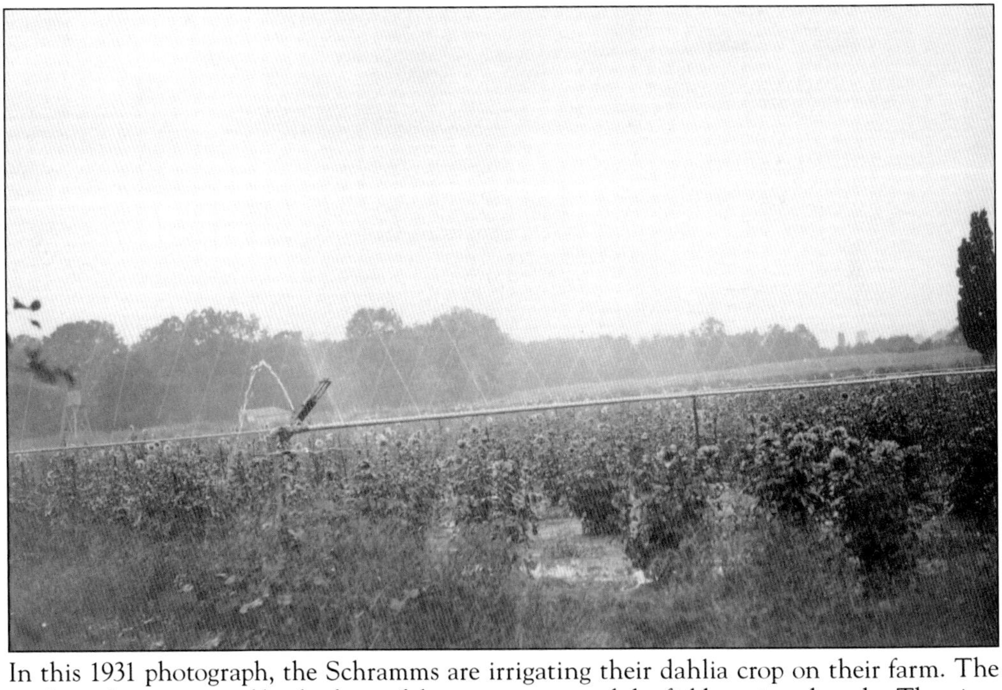

In this 1931 photograph, the Schramms are irrigating their dahlia crop on their farm. The perforated pipes, rotated by the force of the water, protected the fields against drought. The pipes rested on saddles, which were moved every couple of days. Only a few farmers used irrigation at that time. (Schramm.)

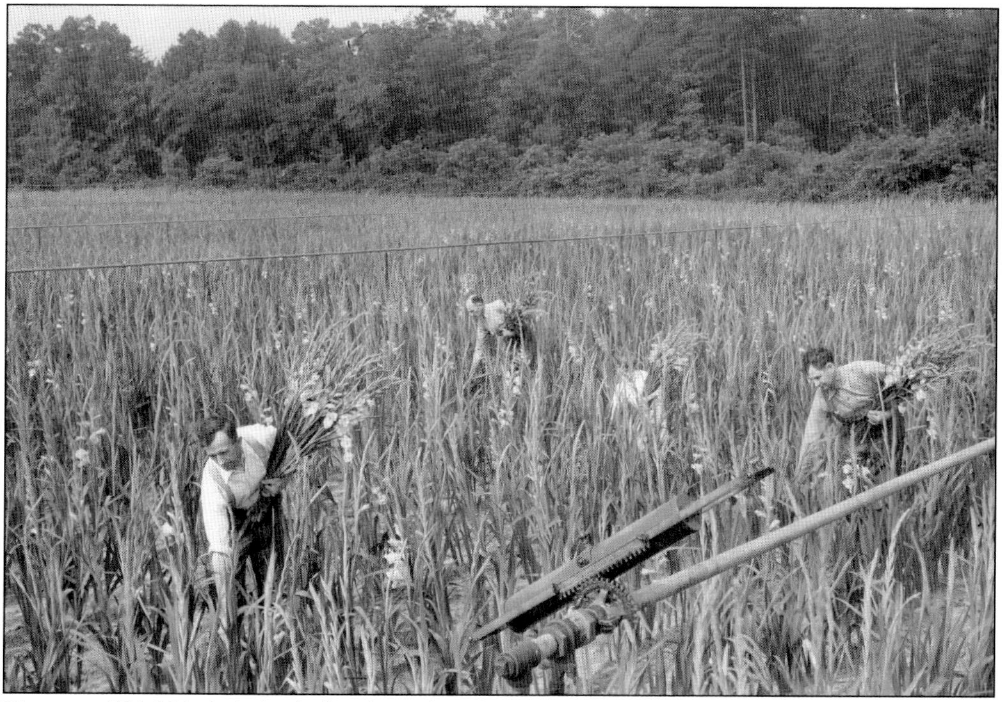

There are 700,000 gladiolus bulbs planted on this 35-acre tract. From left to right, Peter L., Louis Jr., William M., and William M. Schramm II cut the gladiolus flowers to get them ready for market in 1936. William M. Schramm started growing flowers in his greenhouse and fields in 1931. It started as a hobby, but soon, the flowers were making more profit than all other products combined on the farm. (Sun.)

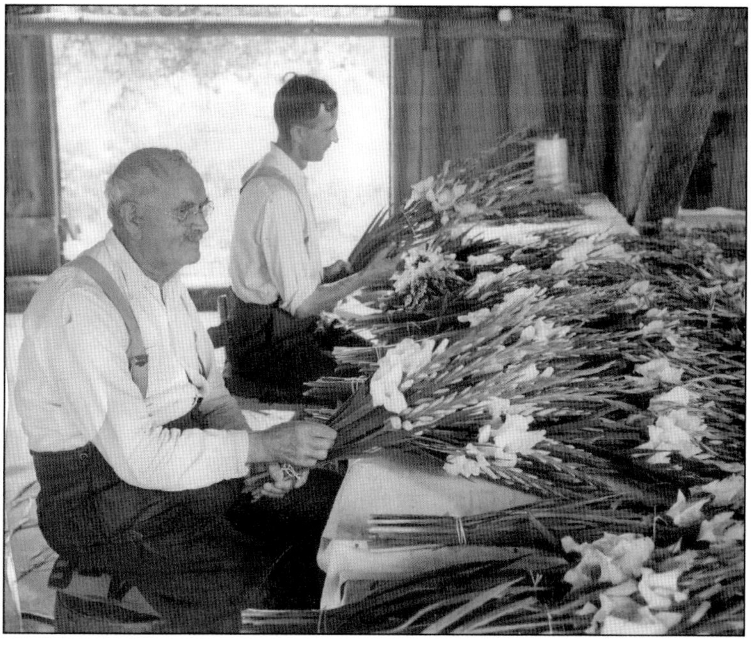

Louis Schramm and one of his sons, Peter L., sort gladiolus blooms for size and color in 1936. Louis Schramm worked at a lime and cement company in Baltimore. Due to health concerns, he was forced to leave his job, so he purchased a farm in Pasadena around 1909. At the height of the cutting season, as many as 8,400 cuttings are harvested and packed for sale in a single day. (Sun.)

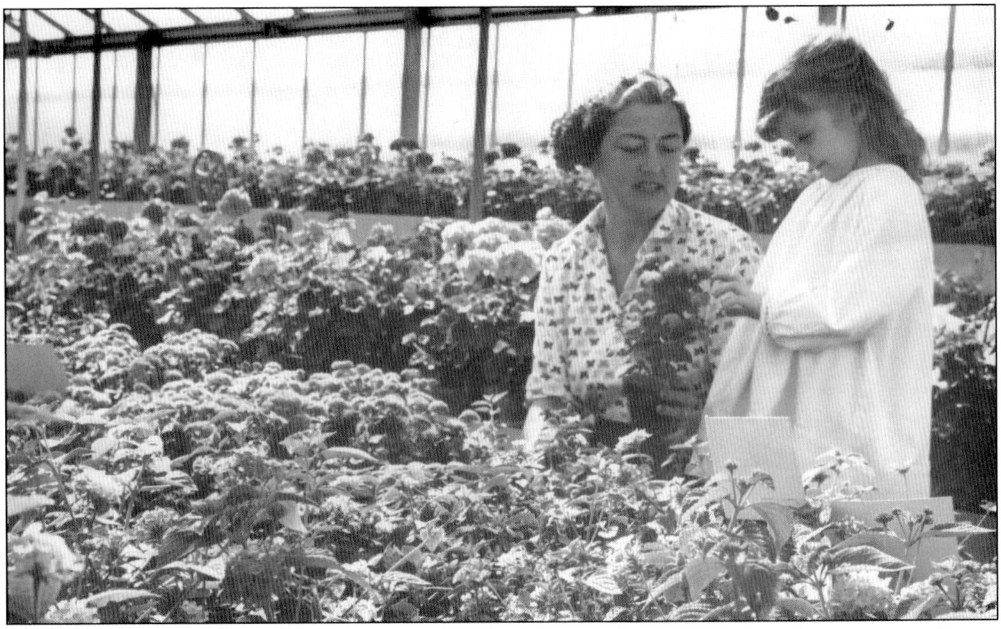

Hildreth Morton loved to share with others her knowledge and passion for growing plants. Heather Ann Drenner listens intently to her grandmother next to a planting bed in one of her glass greenhouses around 1969. Heather worked with Hildreth at Bittersweet Hill Nurseries until Hildreth's death in 2007. (Estate of Hildreth Morton.)

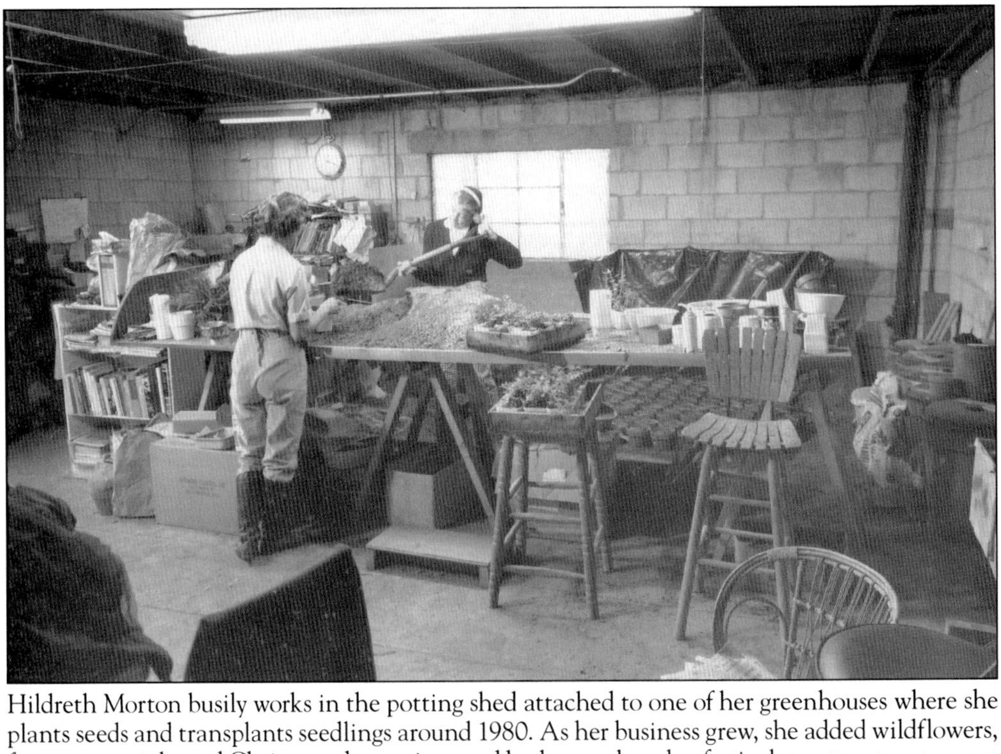

Hildreth Morton busily works in the potting shed attached to one of her greenhouses where she plants seeds and transplants seedlings around 1980. As her business grew, she added wildflowers, ferns, perennials, and Christmas decorations and had annual garden festivals to attract customers. (Estate of Hildreth Morton.)

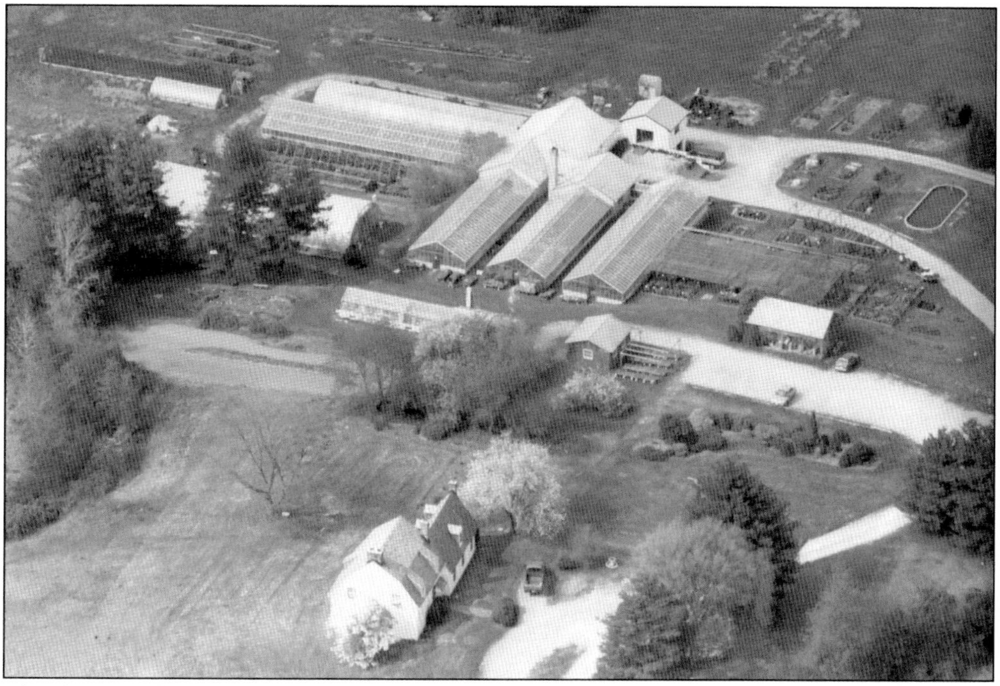

The three Lord Byron English glass greenhouses (center) are seen in this c. 2000 aerial view of the Bittersweet Hill Nursery complex in Davidsonville. Other greenhouses were added as the business grew, along with water lily ponds and a lathe house. The small greenhouse that started her business can be seen in the center of the complex in front of the larger greenhouses. (Estate of Hildreth Morton.)

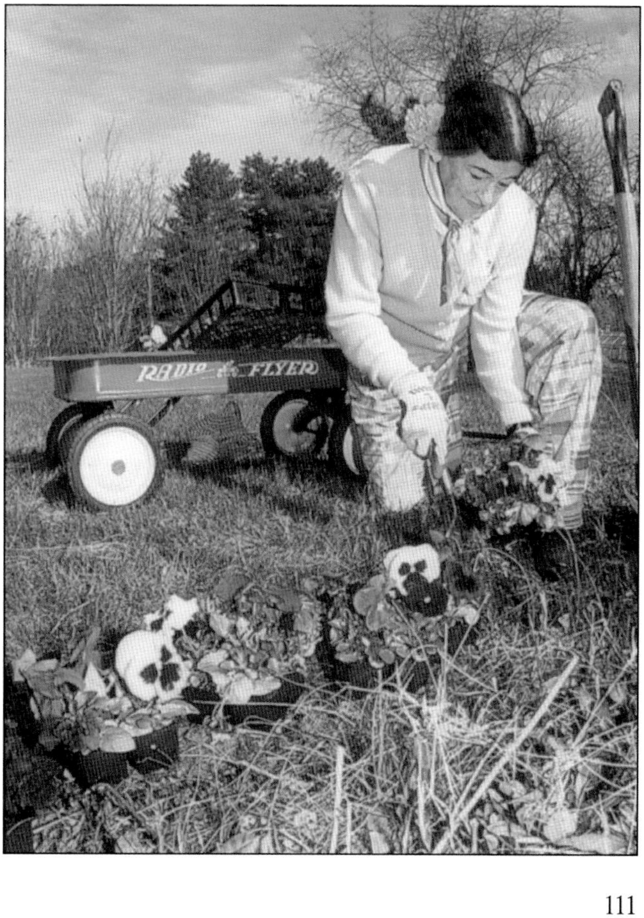

Hildreth Morton plants pansies in one of her display gardens while wearing a fresh flower in her hair in 1999. In an effort to help educate others about horticulture, she was a founding member of the International Water Lily and Water Garden Society and also an active member of the National Herb Society. (Capital.)

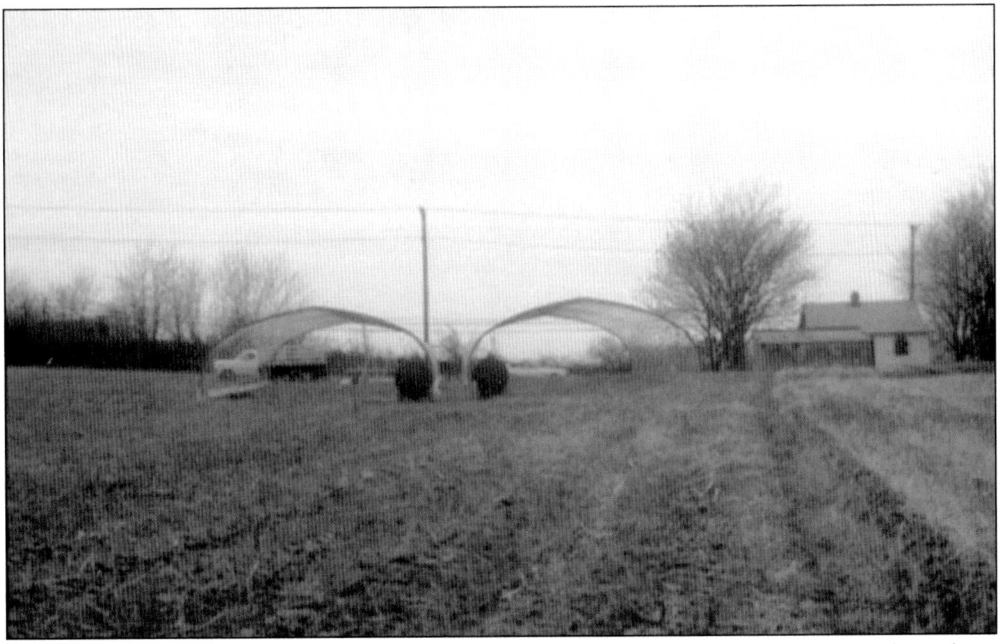

The first two greenhouses are under construction on the future property of Homestead Gardens, pictured looking north toward Route 214 in February 1974. Don E. Riddle Jr. purchased the property in the heart of Davidsonville after outgrowing his original propagation houses in Hyattsville, Maryland. The small white house (right), built at the beginning of the 20th century, served as the administrative office and, in 2010, became the marketing house. (Courtesy of Don Riddle Jr.)

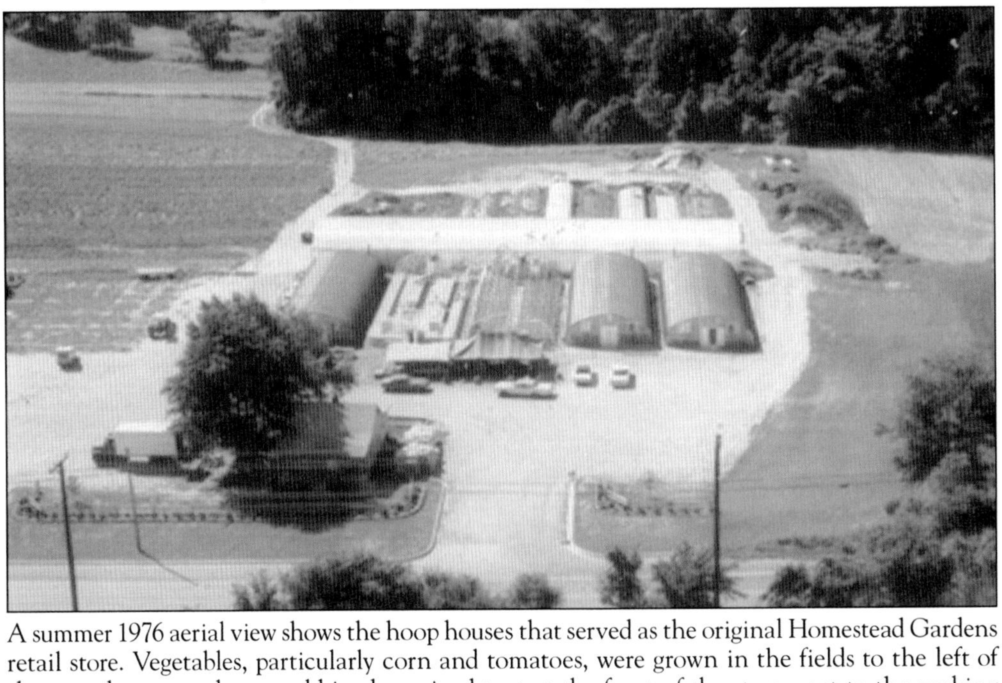

A summer 1976 aerial view shows the hoop houses that served as the original Homestead Gardens retail store. Vegetables, particularly corn and tomatoes, were grown in the fields to the left of the greenhouses and were sold in the striped tent at the front of the store next to the parking area. The administrative offices were located in the house, which stands next to Route 214. (Homestead Gardens.)

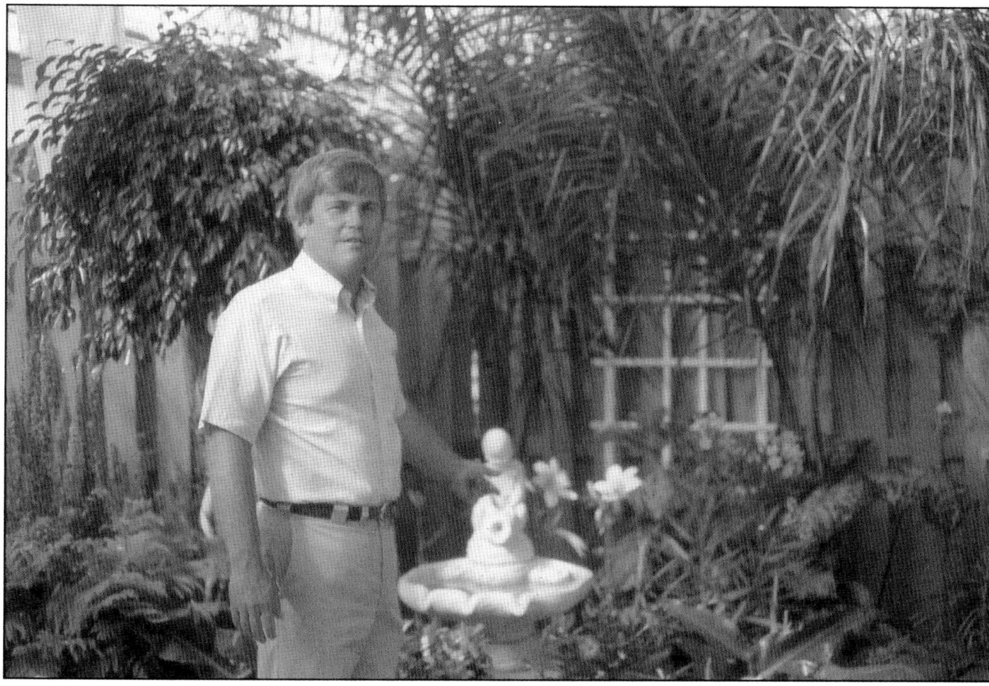

Don E. Riddle Jr. stands in one of the greenhouses at the Homestead Gardens retail store in 1986. In addition to plants, Homestead Gardens sells decorative items for the garden, such as the fountain shown here. (AASCD.)

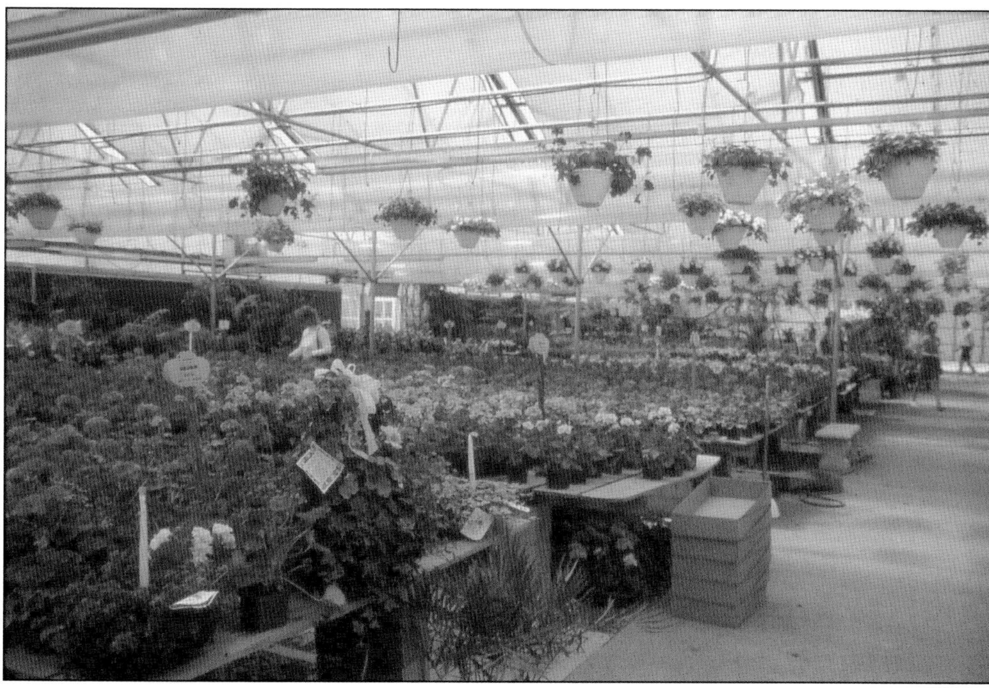

The 1984 interior of the Homestead Gardens retail store shows annuals in pots and in hanging baskets. The glass greenhouses and concrete flooring were installed in 1982 to replace less-solid hoop houses, which had the added benefit of giving a more permanent feel to the store. (AASCD.)

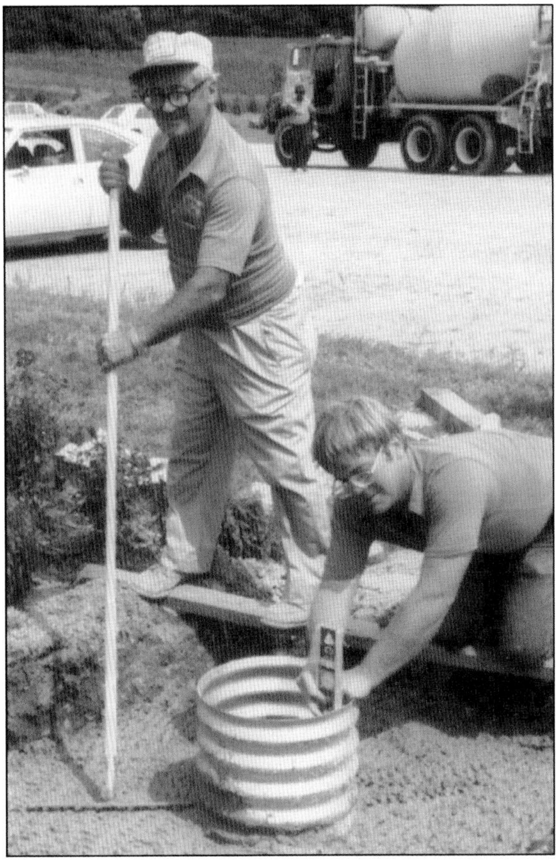

Homestead Gardens founder Don E. Riddle Sr. and Don E. Riddle Jr. install the foundation in 1982 for the 50-foot flagpole that stands at the front of Homestead Gardens. As a US Army veteran of the Normandy invasion, Don Sr. dedicated the flagpole to his comrades who served in World War II. The Military Order of World Wars recognized the Riddle family for their patriotic efforts in 1982. (Homestead Gardens.)

A 1985 aerial view of the Homestead Gardens retail store and growing operation shows the expansion of the business since 1974. Rigid glass greenhouses were built in 1982 to replace the original hoop houses. In 2010, there were 330,000 square feet of protected growing areas. The growing operation on average produces 150,000 vegetable and flower annuals, 85,000 trees and shrubs, and 30,000 poinsettias. (AASCD.)

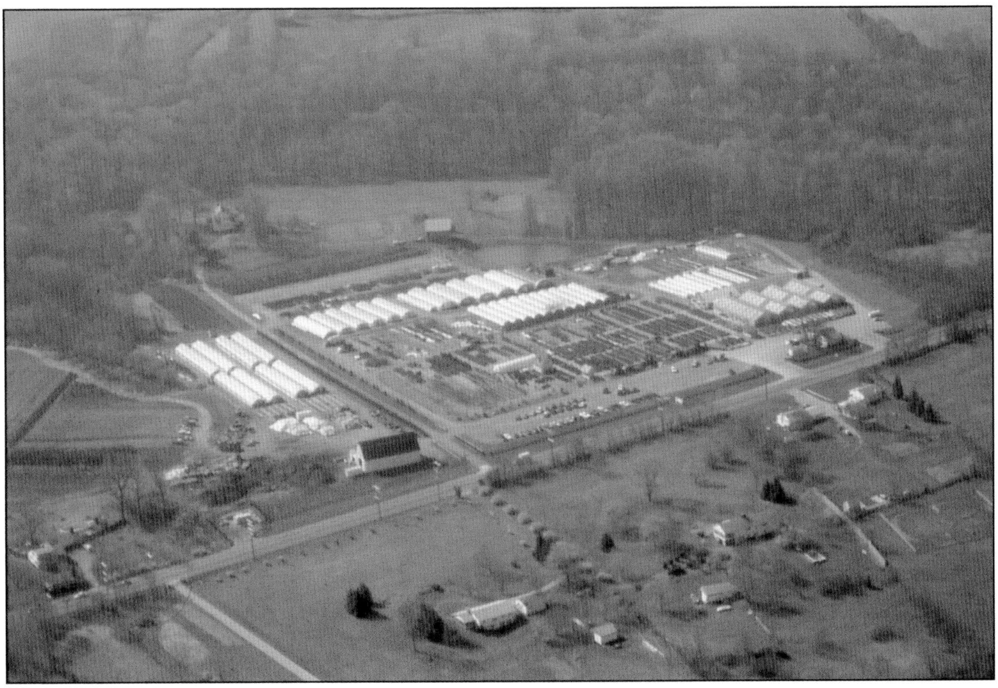

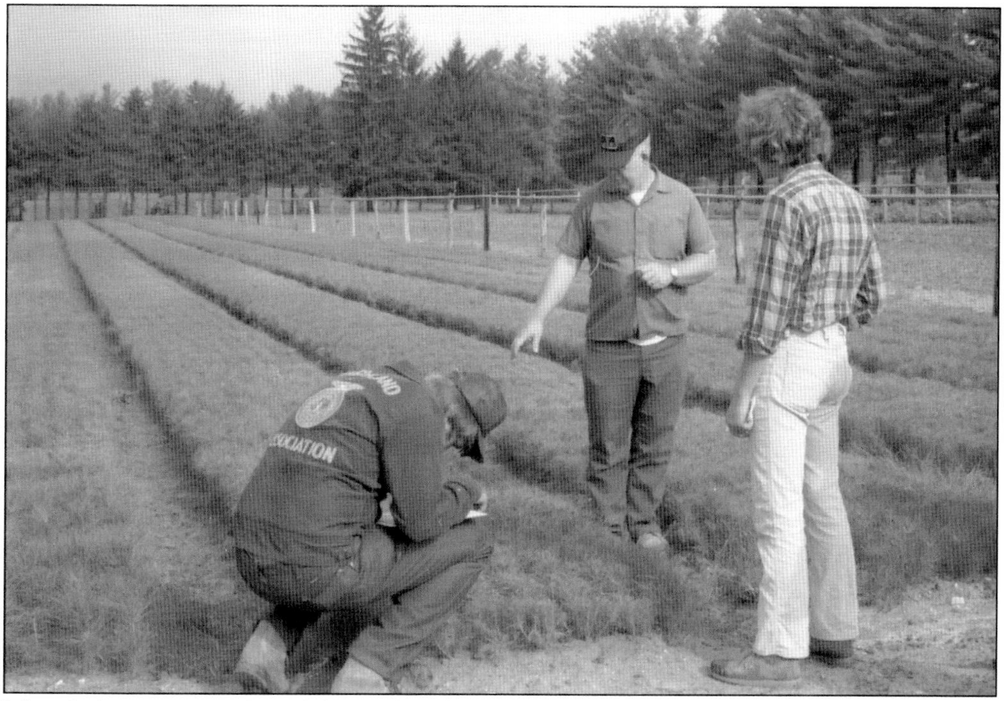

John S. Ayton, supervisor of the Buckingham State Tree Nursery, conducts a tour in 1983 for a group of forestry students from Hereford High School and explains the process for raising superior loblolly and eastern white pine seedlings in the nursery. From 1950 to 1996, the nursery in Harmans had the primary purpose of growing trees for roadside and public plantings. In 1996, the nursery was relocated to the Eastern Shore. (FHD.)

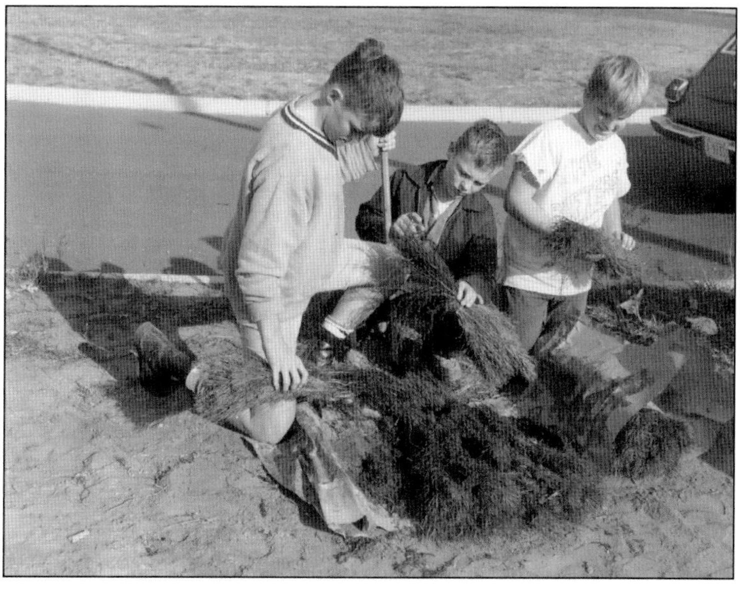

Boy Scouts from Troop 981 examine white pine tree seedlings from the Buckingham Tree Nursery before planting them on the grounds of Providence School in Annapolis in November 1965. This project was a joint venture between the Maryland Department of Forest and Parks, Maryland Cooperative Extension Service 4-H, and the Anne Arundel Soil Conservation District. (AASCD.)

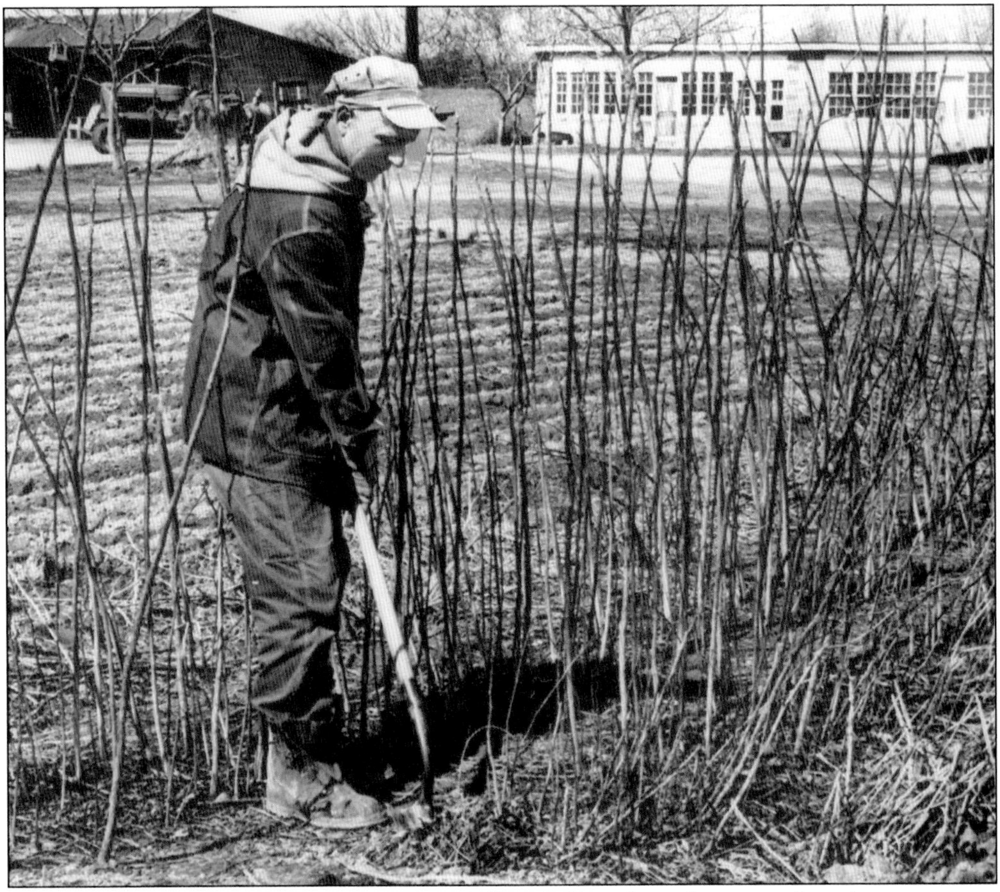

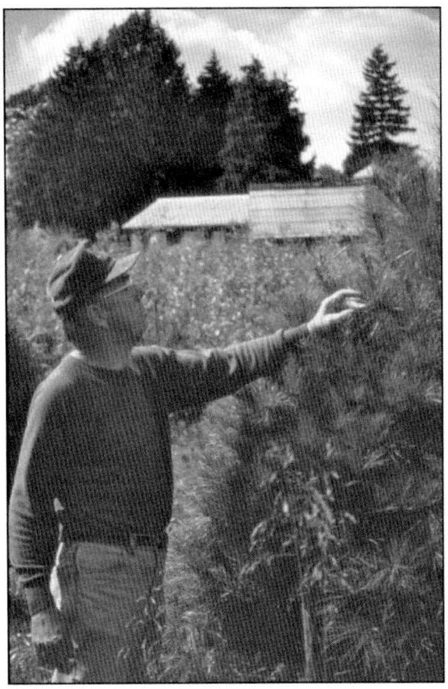

William P. Doepkens digs up saplings he grew from seed or cuttings in his home garden and prepares them to be transplanted in 1969. He was a member of the Anne Arundel County Forestry Board. In cooperation with the local Forestry Board and Soil Conservation Service, since 1961, he planted over 50,000 trees on his farm and the Davidsonville interchange at US Route 50 and Maryland Route 424. (Hearst.)

Richard Smith examines a Christmas tree grown on his tree farm in Harwood in 1989. It takes about 7 to 10 years for a Christmas tree to grow to market size from a seedling. Christmas tree farms and stands are scattered throughout the county. Smith was named Soil Conservation Cooperator of the Year in 1989. (AASCD.)

Six

Managing Our Natural Resources

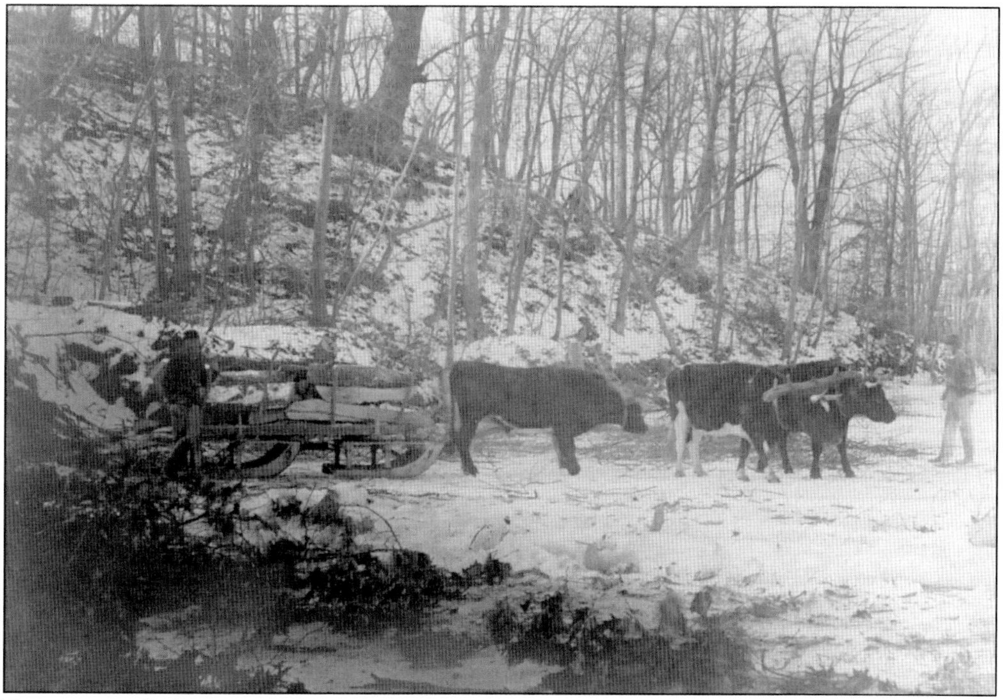

Farmers usually harvest timber in the winter, when the ground is frozen. During a tight freeze, oxen and sleds haul logs across the Rhode River to the Ivy Neck Farm in West River. The timber would be used for building supplies and firewood on the farm or sold. The date of this photograph is unknown. (Colhoun.)

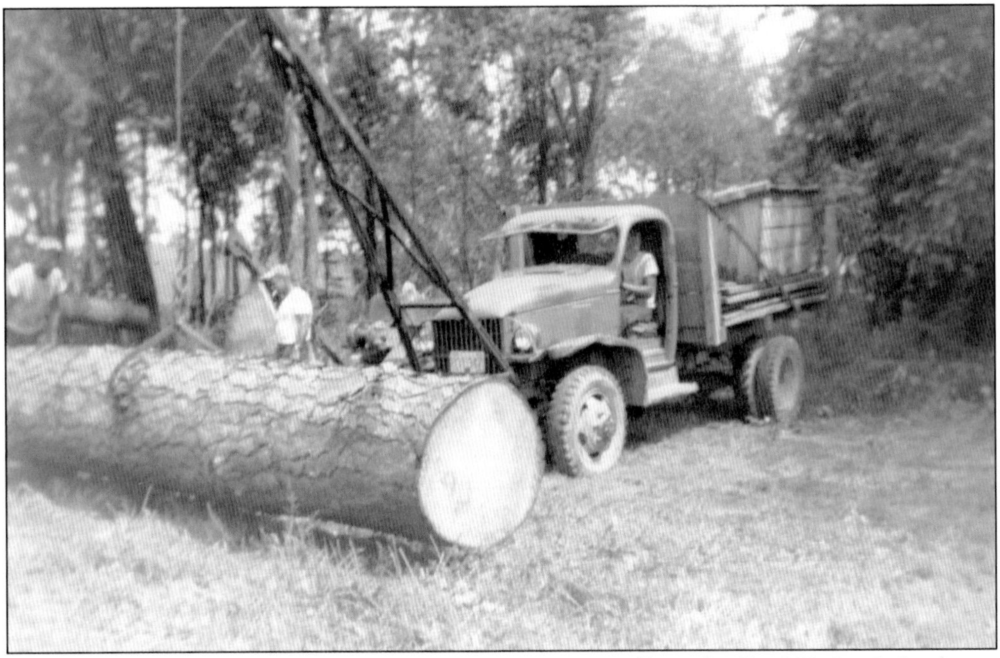

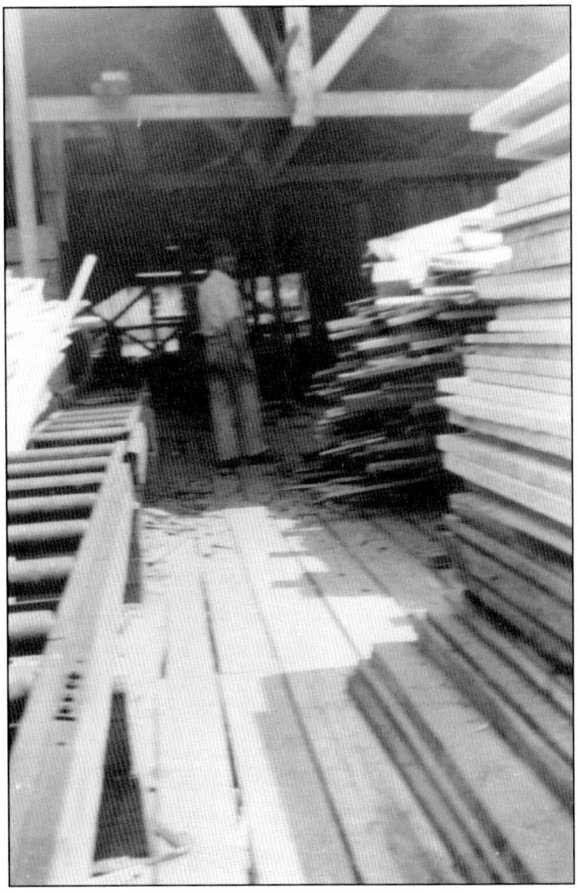

William O. "Boy" Tucker Jr. loads a felled log by means of "Kilroy," a derrick arrangement on the front of a surplus Army truck, around 1940. The log will be put onto a six-by-six truck. Once loaded, it will be hauled to the mill on their farm in Davidsonville. Genie, his wife, used an Allis Chalmers Crawler tractor to help haul logs in their family logging operation. (Chase.)

Boards from sawed logs are stacked for drying at the sawmill on the William O. Tucker Sr. farm in Davidsonville around 1930. Tucker Sr. bought this farm in 1919. During the 1930s, the original farm sawmill was dismantled and replaced by a portable Frick "0-0" mill, gas powered and equipped with a 56-inch circular head saw. During World War II, the sawmill did work for a large lumber company. (Chase.)

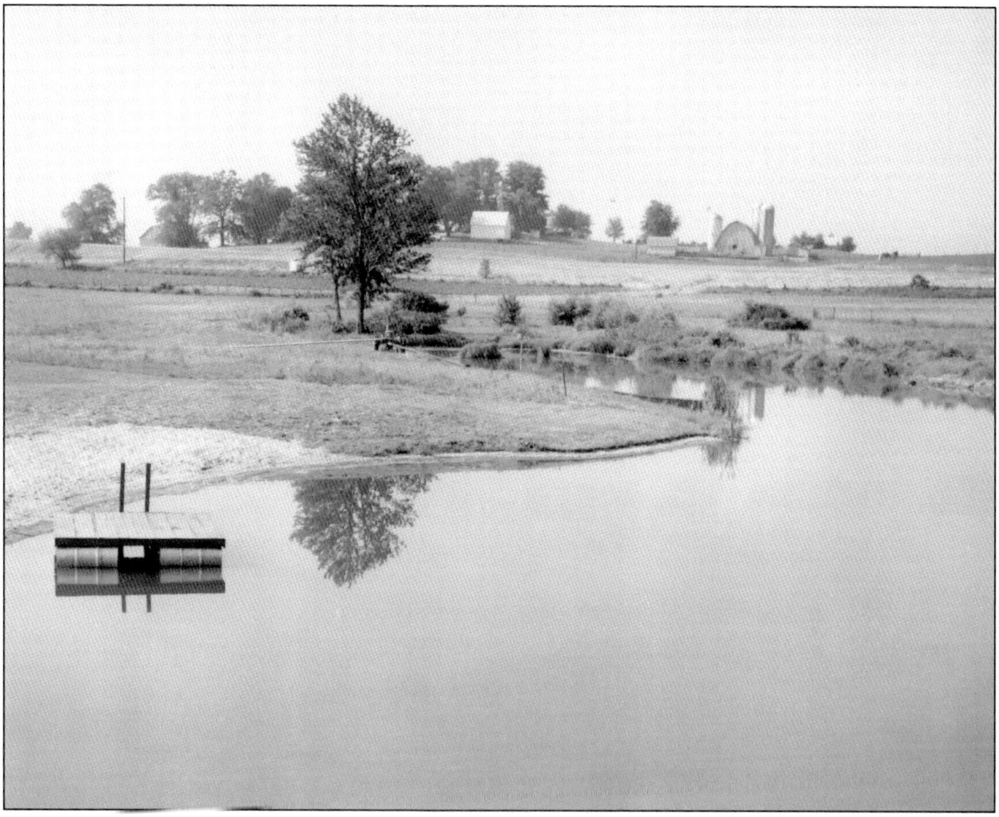

This farm pond on the Anthony Gaug farm in Davidsonville was installed with the help of the Anne Arundel Soil Conservation District. In addition to providing recreation for swimming and fishing and irrigation of farm crops, this farm pond also provided water for firefighters in the community. Benjamin Watkins's dairy farm, along Davidsonville Road, is seen in the background of this May 1963 photograph. (AASCD.)

Anthony Gaug and his sons add fish to their newly completed farm pond in Davidsonville in June 1960. The fish came from the Maryland Department of Game and Inland Fish Commission, which would deliver fingerlings of striped bass and bluegills to farmers in order to stock their farm ponds. Farm ponds were planned with the help of the Soil Conservation Service. (AASCD.)

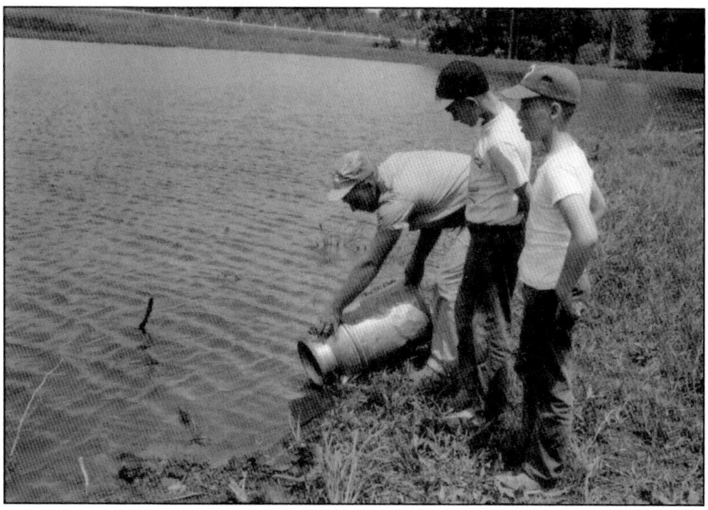

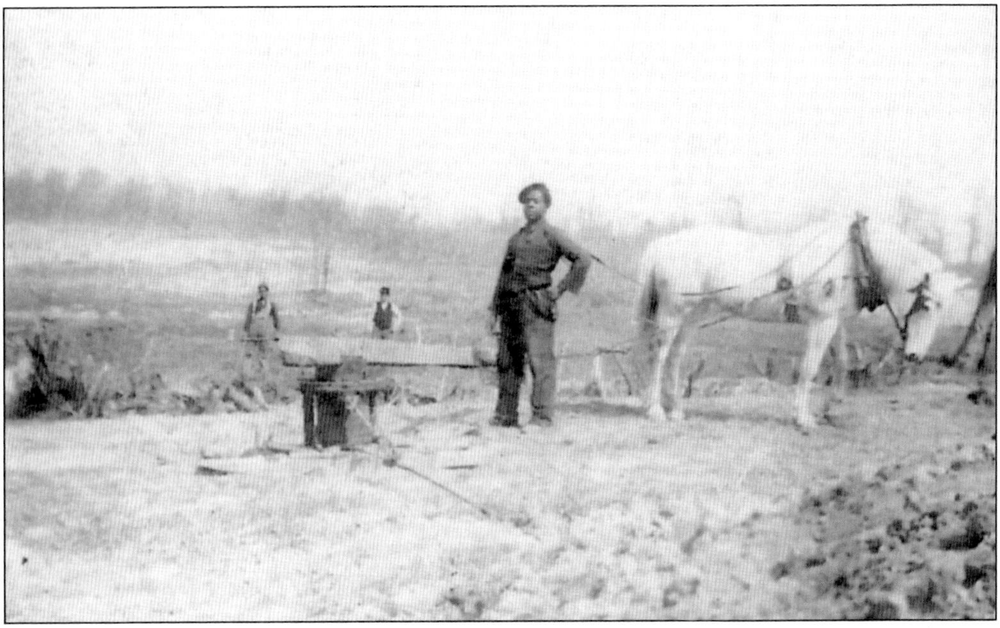

A farm worker uses a stump puller to clear stumps from the Calvert farm pasture around 1930. The stump puller was designed by George W. Calvert, who made patterns to be cast in the Middleton Mead Foundry. It was later used to pull the carriage on the marine railway built on the Calvert farm by George W. Calvert to maintain his power boat and other watercraft owned by his neighbors. (Calvert.)

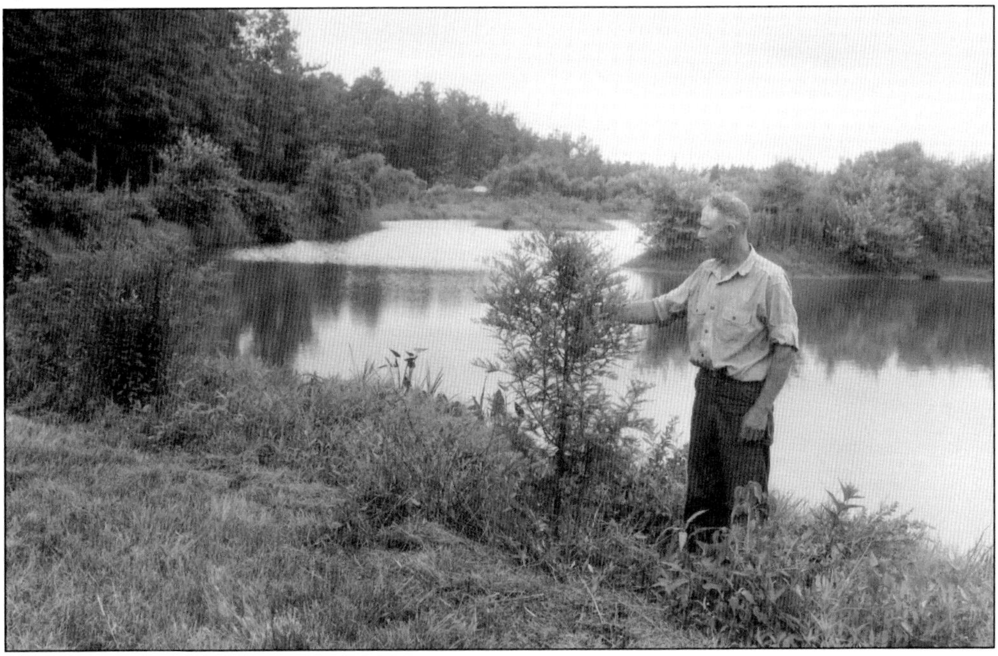

William P. Doepkens, farmer and naturalist, stands in July 1964 next to one of the bald cypress trees he planted around one of the ponds on his farm. Plantings using a variety of trees and shrubs help to provide habitat and food for native species of wildlife. The ponds were installed in the early 1950s with the help of the Soil Conservation Service. (AASCD.)

Seven

Educating our Future

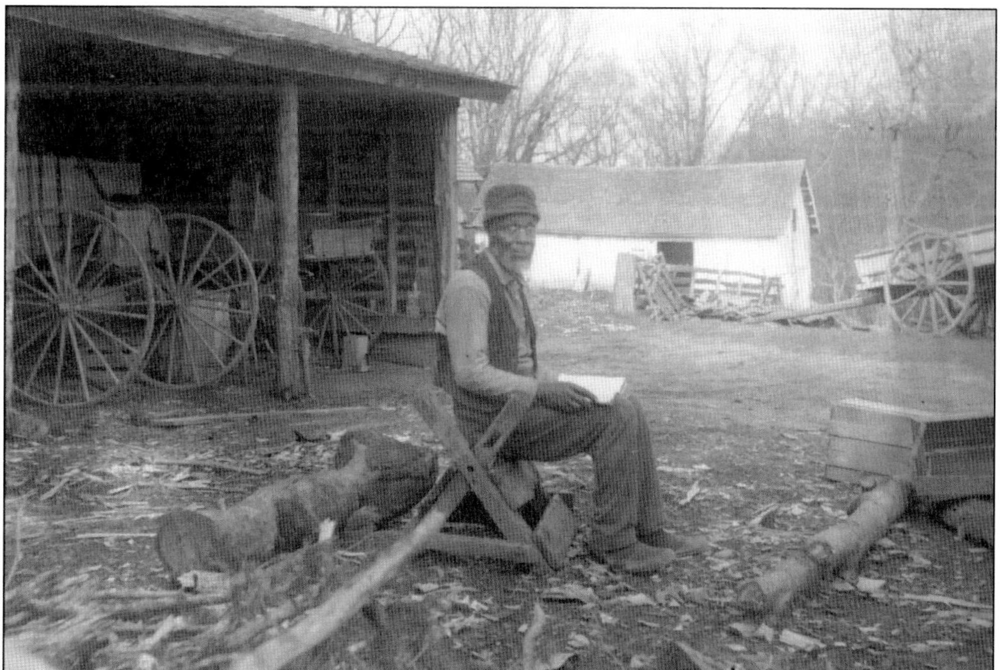

On Ivy Neck Farm, Mr. Dick Pry holds his Bible, which he used to teach himself how to read when he was 60 years old. The date is unknown. (Colhoun.)

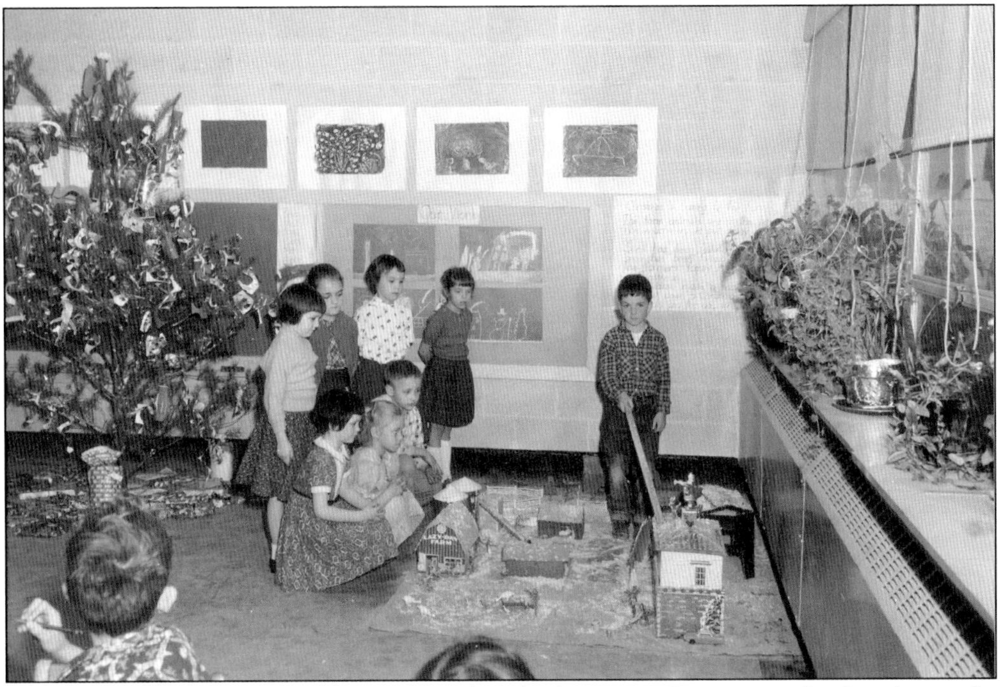

Education begins at all levels. A 1958 second-grade class at an elementary school in Annapolis discusses a conservation project constructed by the class. Since 1989, the Maryland Agriculture Education Foundation has provided education and assistance to schools across the state through various educational programs and teacher workshops. Mobile agriculture classrooms visit schools and fairs in order to teach people about the importance of agriculture. (AASCD.)

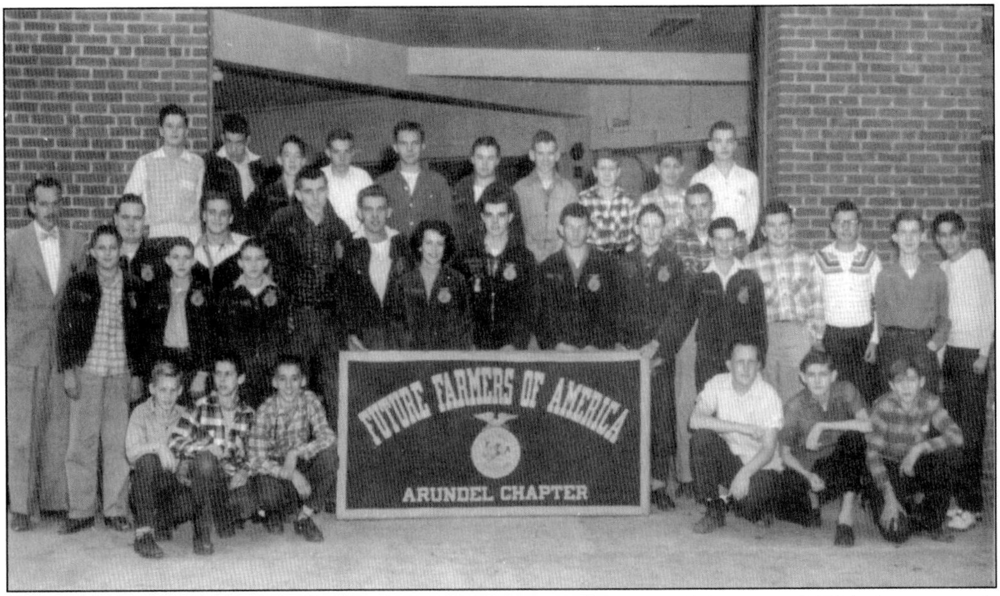

Arundel High School's 1956 yearbook shows the Future Farmers of America (FFA) chapter with their advisor, Jean F. Carlton. The Arundel Agricultural School in Millersville opened its doors in 1924 and provided agricultural education to students at the school until 1957. FFA is a national organization founded in 1928 for students enrolled in agriculture education. (AACPS.)

Maxwell Covington Jr., a member of the Davidsonville 4-H Club, shows off his Grand Champion of Show beef animal at the 1947 Eastern National Livestock Show in Timonium. His parents, Maxwell and Maria Covington, are at his side along with Raymond Heintz (left) and his wife, who purchased the animal for a record-breaking $2.50 per pound. The 4-H is a national organization operated through the Maryland Cooperative Extension Service. (Covington.)

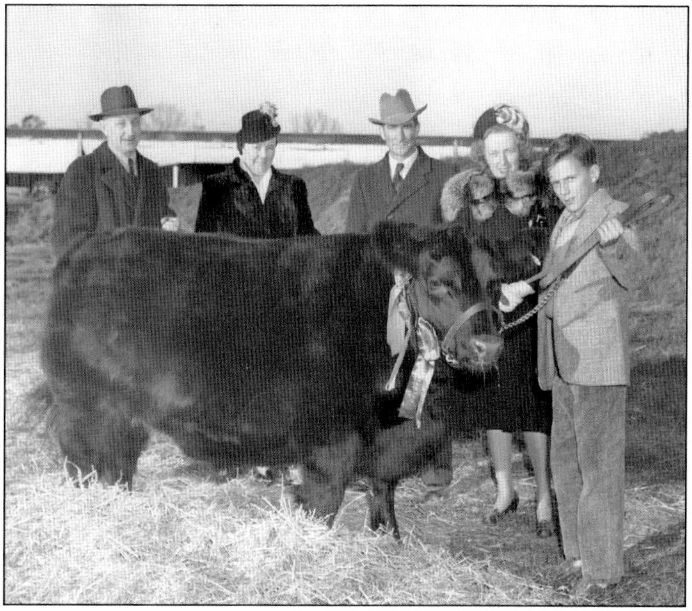

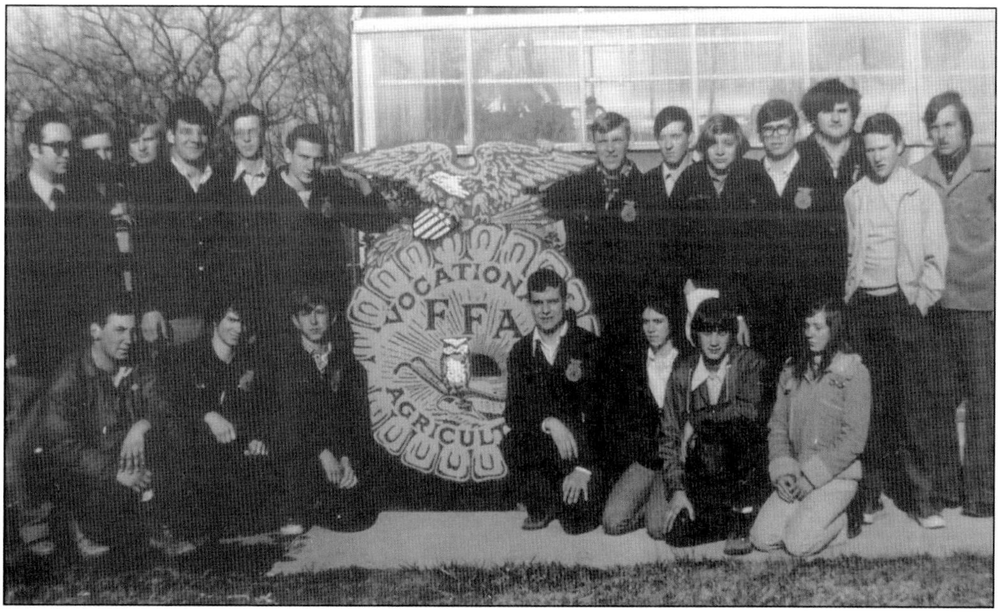

Percy "Putt" Willett (left), horticulture teacher, and Horace Jerry Kennedy (right), agriculture teacher, pose with the 1975 Southern Chapter of the Future Farmers of America in front of the greenhouse at Southern High School in Harwood. Through the inspiration and guidance of their teachers, many of these students went on to become farmers or found employment in other areas of the agricultural industry. (AACPS.)

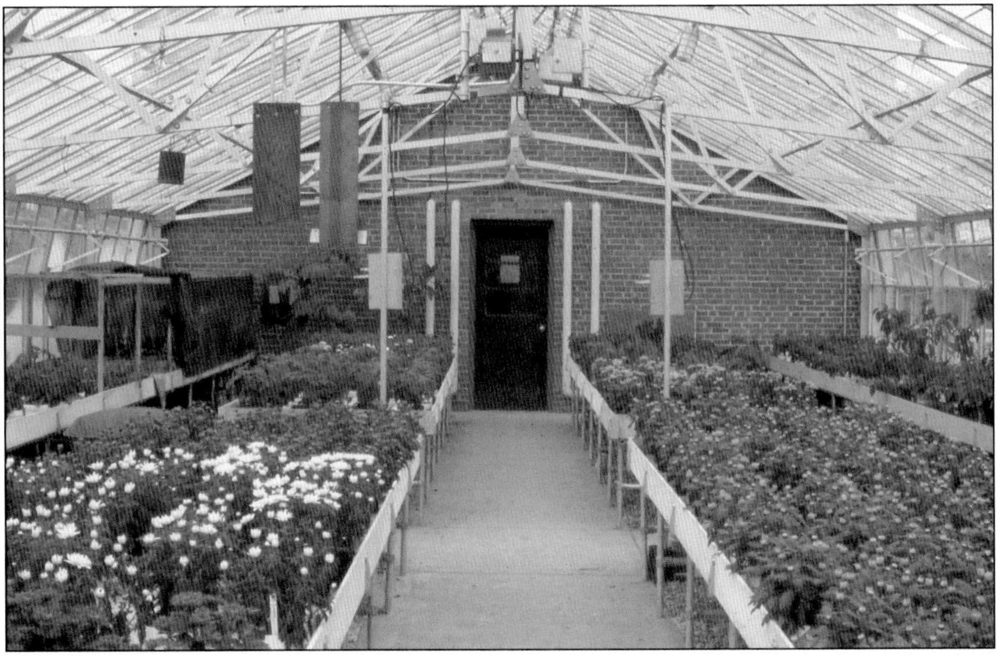

Horticulture students at Southern High School in 1974 used the greenhouse to develop the skills needed to propagate, grow, and market a variety of crops including poinsettias, chrysanthemums, Easter lilies, assorted houseplants, and assorted bedding plants. Sales from the plants were used to help fund the program. (Percy Willett.)

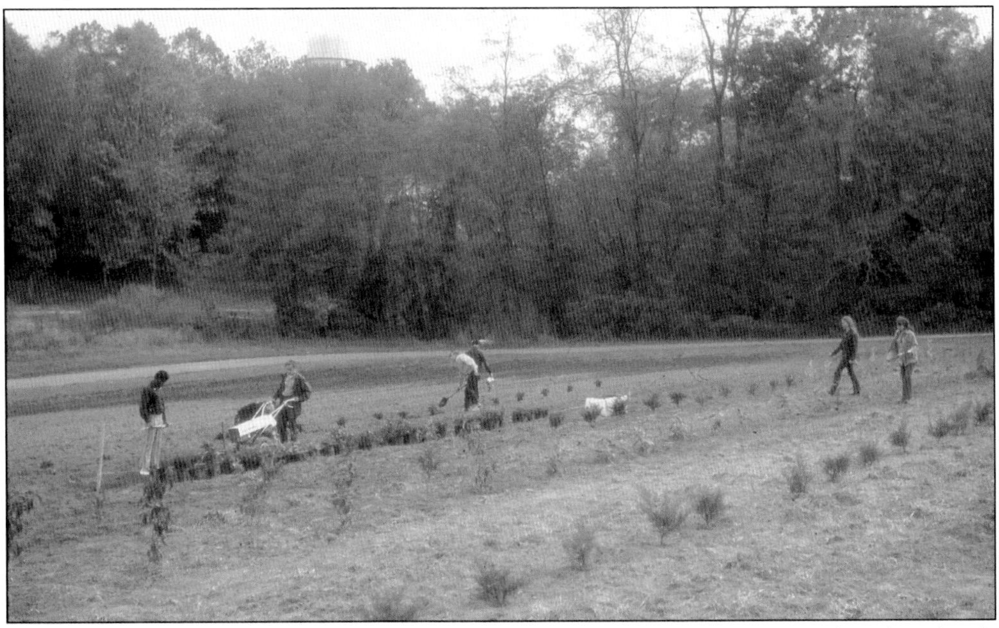

Students enrolled in the horticulture program in 1974 plant and label nursery stock in the school's instructional nursery. The hands-on opportunities provided to the students helped to develop important career skills. The plants were grown to provide shrubs and trees for landscaping projects at schools throughout the county. The program was active from the 1950s until 1983, when it closed; it was reopened in 2010. (Percy Willett.)

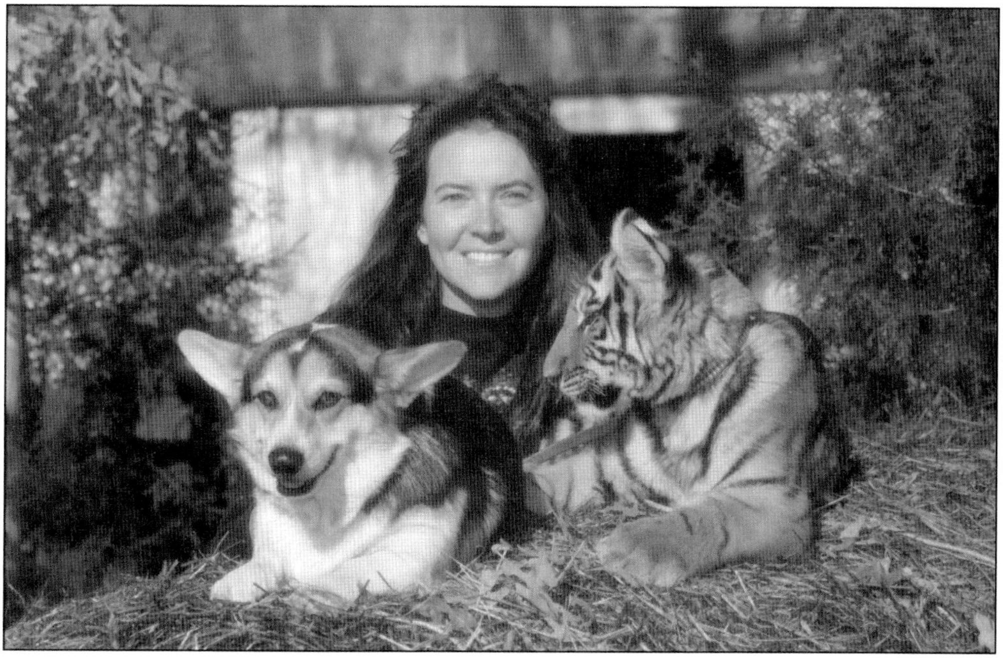

Debra M. Wayson Collison is pictured in 1985 with her dog and Siberian tiger, which she raised from a cub. Steve and Debra Collison established the National Endangered Wild Animal Research & Conservation Center, Inc. (NEWARCC), in 1985 on the Waymore Farm on Governor Bridge Road in Davidsonville. (Debra M. Wayson Collison.)

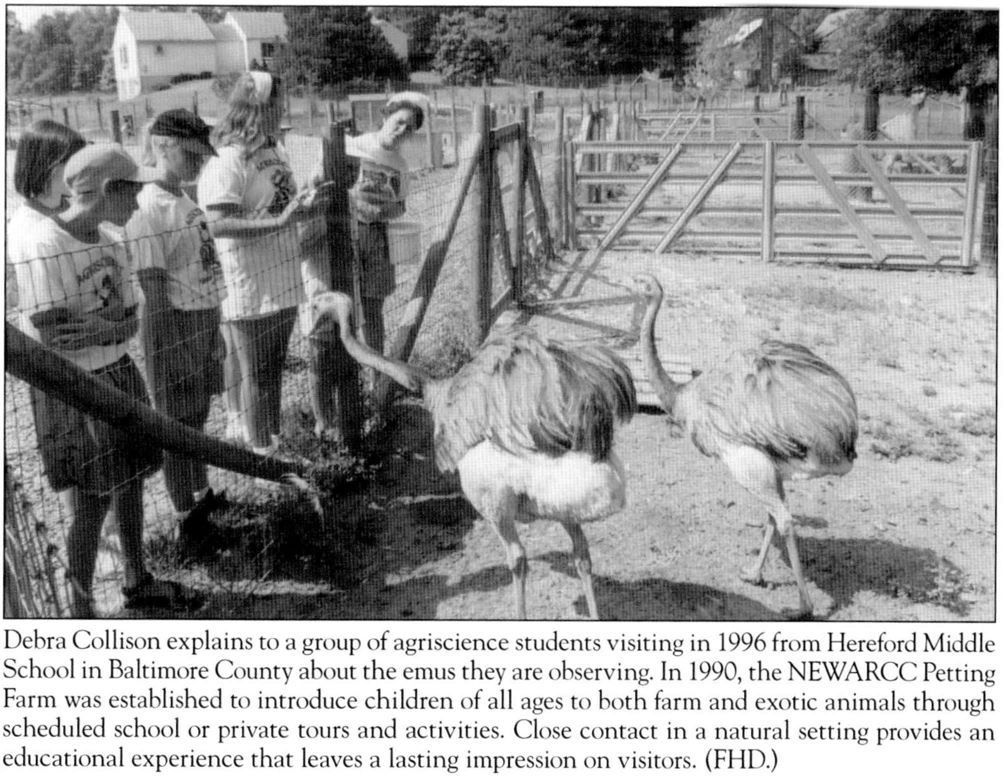

Debra Collison explains to a group of agriscience students visiting in 1996 from Hereford Middle School in Baltimore County about the emus they are observing. In 1990, the NEWARCC Petting Farm was established to introduce children of all ages to both farm and exotic animals through scheduled school or private tours and activities. Close contact in a natural setting provides an educational experience that leaves a lasting impression on visitors. (FHD.)

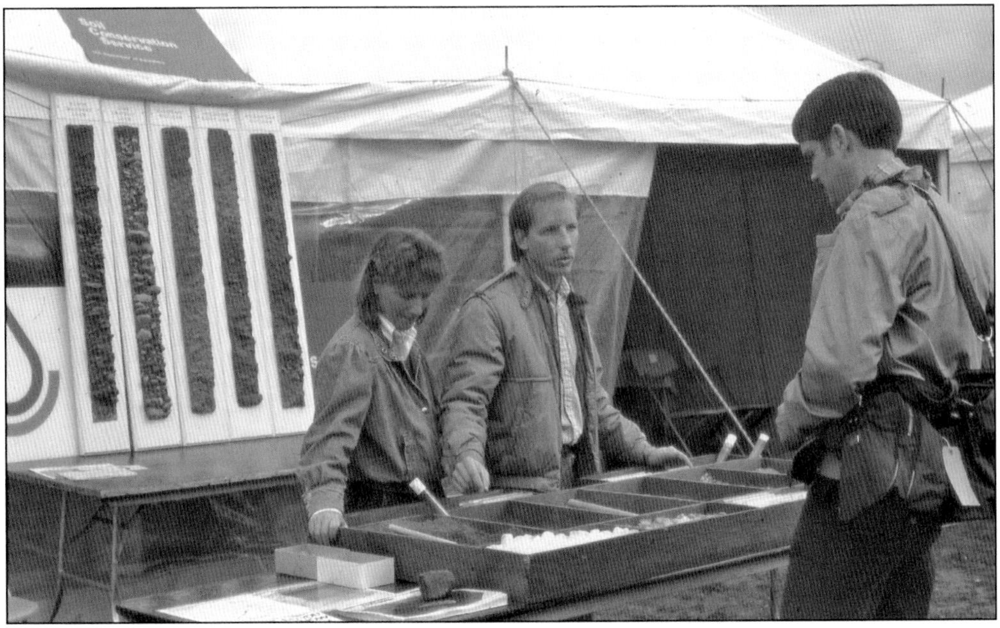

Janet Schmid and John Czajkowski, technicians with the Soil Conservation Service, explain conservation practices at the Chesapeake Bay Appreciation Days held at Sandy Point State Park in October 1986. In addition to public outreach programs, they also conduct regular conservation field days for farmers to promote the latest methods of soil and water conservation practices. (AASCD.)

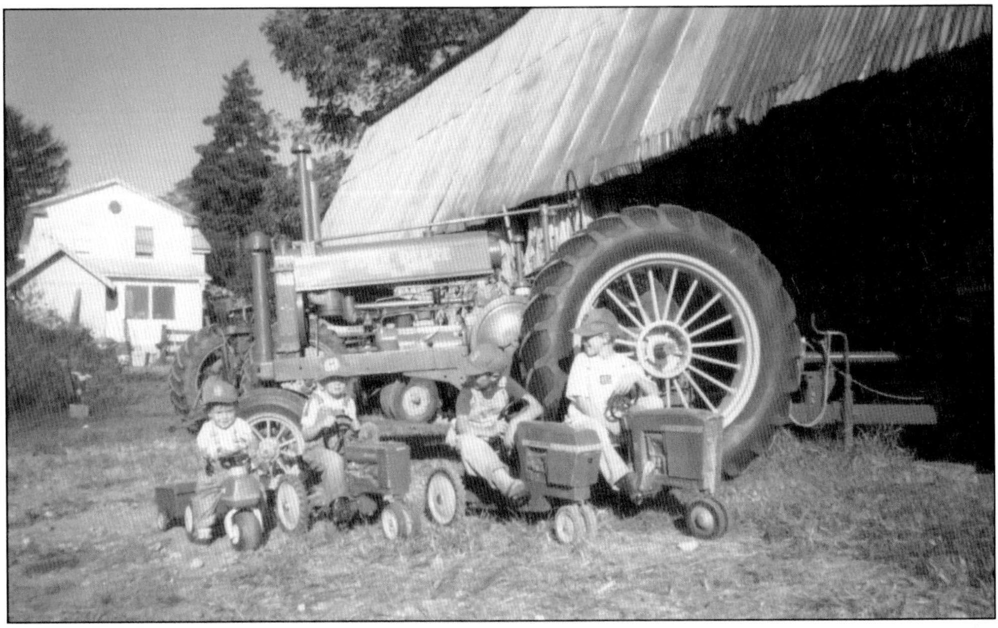

The first John Deere tractor purchased in Anne Arundel County, a 1938 Model A, is seen on the Grimes farm in 2000. In front of the tractor are the children of William Grimes, son of Oscar F. Grimes Jr. The children are, from left to right, Jennylynn, Matthew, David, and John Daniel "J.D." Grimes. In 1986, the farm was put into Farmland Preservation to preserve it as farmland for future generations. (Grimes.)

BIBLIOGRAPHY

Browne, Joseph L. *From Sotweed to Suburbia: A History of the Crofton, Maryland Area, 1660–1960.* Baltimore: Gateway Press Inc., 1985.

Doepkens, William P. *Excavations at Mareen Duvall's Middle Plantation of South River Hundred.* Baltimore: Gateway Press Inc., 1991.

Mumford, Willard R. *Strawberries, Peas & Beans—Truck Farming in Anne Arundel County.* Linthicum, MD: Ann Arundel Historical Society, 2000.

The Tobacco Institute. *Maryland & Tobacco: A Chapter in America's Industrial Growth.* Washington, DC: The Tobacco Institute, 1976.

US Department of Agriculture. *The Census of Agriculture.* National Agricultural Statistics Service, www.agcensus.usda.gov.

Consistent with our mission to preserve history on a local level, this book was printed in South Carolina on American-made paper and manufactured entirely in the United States. Products carrying the accredited Forest Stewardship Council (FSC) label are printed on 100 percent FSC-certified paper.

MADE IN THE USA